TRANSFORMING HOBBY DRONES FOR WARFARE

PETER SAVAGE

Publication Disclaimer

This book is intended for educational and historical purposes only, offering readers an informative overview of the development and evolution of drone technology. It explores the connection between significant historical events and the advancement of unmanned aerial systems (UAS), providing context for how drones have influenced military, scientific, commercial, and recreational sectors. The content is based on publicly available information, engineering knowledge, and historical records, aiming to deepen understanding of the milestones that have shaped modern aerial robotics.

Drones, artificial intelligence, and machine learning have fundamentally redefined warfare, marking the greatest transformation in conflict since gunpowder reshaped the battlefield.

Readers must not interpret this material as instruction or endorsement for building, modifying, or operating drones in ways that breach local, national, or international laws. The construction or use of unauthorized or weaponised drones—particularly for surveillance or combat—is expressly discouraged. The authors and publishers do not support any unlawful applications of the information. Drone regulations differ across jurisdictions and are subject to frequent changes. It is each individual's responsibility to remain informed about current legal and safety requirements concerning drone ownership and operation. This includes compliance with rules on airspace, registration, payloads, and flight conduct. This book promotes the ethical use of engineering knowledge and encourages responsible engagement with drone technology, underscoring the importance of legality, safety, and innovation

Other books in the Putin's Pathway series

Ukraine, The Good, Bad and the Ugly, From the Bucha Massacre, Liberation of Kherson to the fall of Bakhmut

Part Time Soldier, Full Time Warrior

ISBN: 978-1-7643594-3-6 E Pub

ISBN: 978-1-7643594-4-3 E Book

ISBN: 978-1-7643594-5-0 Paperback - Illustrations, Full Colour

ISBN: 978-1-7643594-6-7 Paperback - Illustrations Black & White

Preface

In the early days of the 2022 full Russian invasion of Ukraine, during the initial offensive aimed at capturing Kyiv, Peter Savage travelled to the region in his capacity as a Chaplain. His mission was to assist in coordinating humanitarian aid from Australia and the United Kingdom for severely affected communities and those critically wounded by the conflict.

What he encountered was far beyond what most could imagine. Among the most harrowing experiences was his visit to Bucha, a town that had recently endured a horrific massacre during the battle in February. There, Peter witnessed the aftermath of atrocities committed against civilians—families, children, and infants—many of whom had sought refuge in the basements of residential buildings. Victims had been found handcuffed, executed with shots to the head, and their bodies burned in what appeared to be an attempt to conceal evidence of war crimes.

It is difficult to convey the emotional weight of what Peter experienced. Among the most haunting images were those of mothers cradling their children in death, their bodies intertwined—both shot and burned—preserved only as charred reminders of unimaginable suffering.

Motivated by what he had seen, Peter wrote a short account for publication and remained in Ukraine, continuing his work within the aid community. He formed a close connection with St Sophia Cathedral in Kyiv and worked to support both spiritual and material relief efforts. However, his background in electronics, microprocessors, and military service—including service in NATO and the Air Force—soon proved invaluable in unexpected ways.

As Peter travelled deeper into the conflict zones, including Kherson, Bakhmut, and the Donbas region, he began observing a significant shift in how civilians and soldiers were adapting to the evolving nature of warfare. In particular, he noted the innovative and widespread use of hobby drones—initially designed for recreation—re-purposed for front-line military applications such as reconnaissance and improvised bombing missions using hand grenades.

Though his primary role was humanitarian, Peter paid close attention to the grassroots ingenuity around him. The repeated requests he heard from Ukrainian fighters and volunteers were not only for food, medicine, and protective gear, but also for drones, spare parts, and 3D printers. These tools were urgently needed to modify and manufacture components for the ongoing adaptation of consumer drones into tactical battlefield assets.

Through extensive note-taking, interviews, and firsthand observation, Peter developed an in-depth understanding of these grassroots efforts and the emerging role of artificial intelligence in a new and complex form of warfare. He had the opportunity to meet with several battalion commanders who shared their insights into how drones had rapidly become indispensable tools in this evolving conflict.

Shipments of hobbyist equipment—ranging from balsa wood fixed-wing aircraft kits to rotary drone motors and 3D-printed components—were now flowing into Ukraine from suppliers across Europe and beyond. What was unfolding before Peter's eyes was not just a response to necessity, but the birth of a radically new mode of warfare: one in which civilian-grade technologies and off-the-shelf components were being re-purposed at scale for lethal effectiveness on the modern battlefield.

This book presents a window into Peter's journey—what he saw, what he learned, and the profound ethical questions that arise from the militarisation of civilian technology in a time of total war. This book is a first-hand account of how hobby drones evolved into weapons of war, witnessed on the ground as necessity, innovation, and survival reshaped the modern battlefield.

Contents

Appendix(s)

CHAPTER 1

Introduction, Hobby Drones for Warfare

1.1 Hobby Drones for Warfare

Drones have emerged as a revolutionary tool in modern warfare, significantly altering the dynamics of conflict and military strategy. Initially developed for recreational purposes, hobby drones have been adapted for tactical use, showcasing their versatility and potential on the battlefield. This adaptation has led to a new wave of interest among enthusiasts and military personnel alike, as the lines between civilian technology and military applications continue to blur. Understanding this transformation is crucial for those interested in the evolving landscape of drone technology and its implications for warfare.

The integration of rotary and fixed-wing unmanned aerial vehicles (UAVS) into military operations has opened new avenues for reconnaissance, surveillance, and direct engagement. Rotary drones, known for their vertical take-off and landing capabilities, are particularly effective in urban environments where agility and maneuverability are paramount. Conversely, fixed-wing drones offer extended flight times and broader operational ranges, making them ideal for extended missions that require sustained surveillance. This book will explore the distinct advantages of each type of UAV, emphasising how hobbyists can leverage their understanding of these systems for military applications.

DIY modifications play a critical role in enhancing the durability and effectiveness of drones in combat scenarios. Hobbyists often possess the skills and creativity necessary to make significant upgrades to off-the-shelf models, improving their resilience against harsh conditions and enemy detection. By examining various modification techniques, the publication will also provide practical insights into how enthusiasts can optimise their drones for tactical use, including reinforcing frames, improving battery life, and upgrading sensors for enhanced performance.

Autonomous navigation systems represent a significant advancement in drone technology, allowing for precise and efficient tactical operations. These systems enable drones to operate with minimal human intervention, reducing the risk to operators and enhancing mission effectiveness. The development and integration of AI and machine learning algorithms further enhance autonomous capabilities, allowing drones to adapt to dynamic environments and make real-time decisions. This discussion will highlight the importance of these technologies in modern warfare and how hobbyists can explore their implementation in modified UAVs.

Throughout this book, the ethical implications of adapting hobby-grade drones for military use are critically examined against the harsh realities of total war. While an understanding of military regulations, international humanitarian law, and domestic legal frameworks is essential for anyone navigating this complex terrain, history demonstrates that the pace of conflict often outstrips the speed of regulation. In high-intensity warfare—particularly when facing an aggressive adversary with superior resources—states and non-state actors alike may prioritise operational survival over procedural compliance. This tension between legal restraint and battlefield necessity forms one of the central dilemmas explored in these pages. The rapid evolution of unmanned systems illustrates how innovation, driven by urgency, can both challenge and reshape existing legal and ethical boundaries.

It is essential to recognise that the imperative to modify hobby drones for warfare frequently supersedes both industrial constraints and ethical considerations governing their use. Although international regulations prescribe standards for the manufacture and deployment of unmanned aerial systems (UAS), advancements in technologies beyond these regulatory frameworks have led to the adoption of alternative production methods and the utilisation of previously restricted communication protocols, frequency bands, and network architectures.

Moreover, there is growing concern over the use of indiscriminate or previously banned ordnance, such as cluster munitions or thermobaric weapons, when deployed via drones. These modified UAVs can be used in swarming tactics, saturating an area with explosives, regardless of civilian presence. Without apparent oversight and in the absence of enforceable norms, drones become tools not only of strategic precision but of ethical ambiguity and, at times, outright lawlessness. As the technological ceiling continues to rise and the barriers to drone modification continue to lower, the international community must grapple with the reality that the nature of war is changing faster than the legal governance.

1.2 Quad-Copter and Rotary UAVs

Rotary Unmanned Aerial Vehicles (UAVs), commonly known as drones, have gained prominence in both civilian and military applications due to their versatility and manoeuvrability. These UAVs, characterised by their rotating blades, can take off and land vertically, making them ideal for operations in confined spaces and challenging terrains. Their ability to hover enables detailed reconnaissance and surveillance missions, which are critical in tactical situations where ground visibility is limited. The design of rotary UAVs enables them to carry various payloads, including cameras, sensors, and even small munitions, enhancing their role in modern warfare.

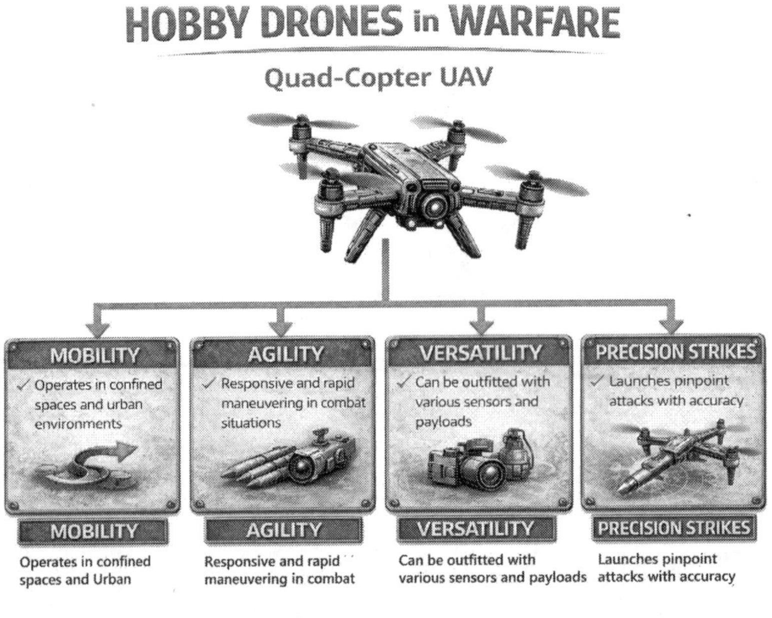

The transformation of hobby drones into tactical rotary UAVs involves significant modifications to improve their durability and functionality in combat scenarios. Enthusiasts often start with commercially available models, integrating robust materials and components to withstand harsh environments. Enhancements may include improved battery life for extended flight times, reinforced frames to withstand impacts, and weather-resistant features that enable operation in adverse conditions. Through DIY modifications, hobbyists can create drones that are not only resilient but also capable of performing multiple missions, from intelligence gathering to delivering supplies.

Autonomous navigation systems play a crucial role in the effectiveness of rotary UAVs during tactical operations. These systems enable drones to operate with minimal human intervention, allowing for pre-programmed flight paths and real-time adjustments based on environmental variables. Advanced algorithms and sensor integration allow rotary UAVs to navigate complex urban landscapes, avoiding obstacles and maintaining stable flight. By incorporating artificial intelligence and machine learning, these drones can analyse data in real-time, making decisions that enhance mission success rates while reducing the risk to operators.

Payload optimisation techniques are also essential for maximising the effectiveness of rotary UAVs in surveillance and targeting roles. Understanding how to balance the weight of additional sensors or munitions with the drone's flight capabilities is crucial. Enthusiasts can experiment with different configurations to ensure that their UAVs can carry the necessary equipment without sacrificing flight performance. This involves not only optimising the payload but also ensuring that the drone's communication systems are secure, allowing for uninterrupted data transmission during critical missions.

The integration of rotary unmanned aerial vehicles (UAVs) into military operations raises important ethical considerations, particularly regarding their use in warfare. While these modified drones offer tactical advantages, their deployment should be carefully regulated to prevent misuse and ensure compliance with legal frameworks. Hobbyists and military personnel alike must be aware of the implications of using drones in combat, including the potential for collateral damage and the moral responsibilities that come with deploying aerial surveillance and attack capabilities. Balancing innovation with ethical considerations is vital as the landscape of drone warfare continues to evolve.

1.3 Fixed-Wing UAVs

Fixed-wing UAVs, often characterised by their aeroplane-like design, offer unique advantages in terms of endurance, speed, and range compared to their rotary counterparts. These attributes make them particularly suitable for a variety of tactical applications, especially in military contexts where prolonged light duration and the ability to cover large areas are critical. Enthusiasts looking to modify hobby drones for tactical use can benefit from understanding the capabilities inherent in fixed-wing designs, allowing for effective reconnaissance and intelligence-gathering missions.

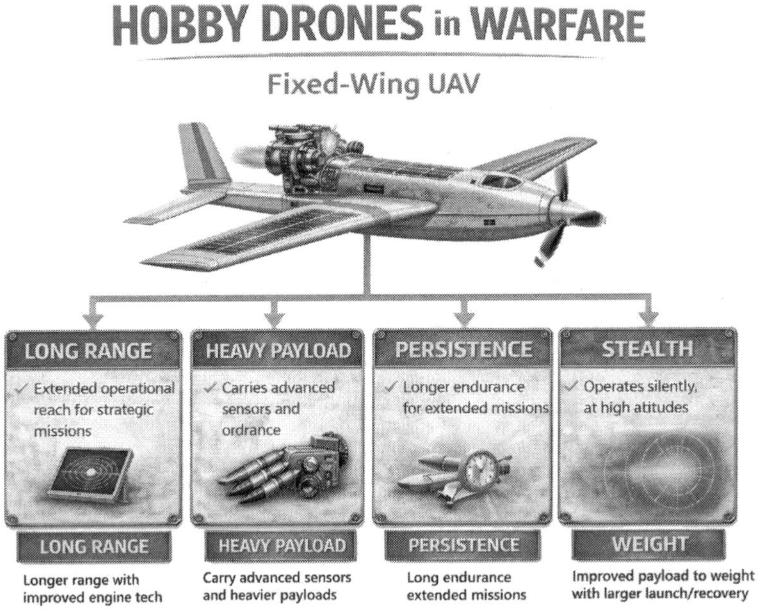

One of the primary benefits of fixed-wing UAVs is their ability to achieve greater flight efficiency. The aerodynamic structure allows these drones to glide for extended periods, significantly increasing their operational time in the field. This endurance is vital for missions that require continuous surveillance or data collection without the need for frequent recharging or battery swaps. Hobbyist modifications can further enhance this flight time by optimising weight distribution and incorporating lightweight materials, ensuring the UAV remains agile while maximising its payload capacity.

The integration of autonomous navigation systems is another area where fixed-wing UAVs can be transformed for tactical use. These systems enable drones to follow pre-planned flight paths, making them ideal for automated surveillance missions. Enthusiasts can explore DIY modifications that incorporate advanced GPS and inertial navigation systems, allowing for higher precision in target tracking and data collection. This not only enhances the operational capabilities of hobby drones but also aligns them with military standards, making them useful in various tactical scenarios.

Payload optimisation is crucial for maximising the effectiveness of fixed-wing UAVs in military applications. Modifications can include the installation of high-resolution cameras, thermal imaging sensors, or even electronic warfare equipment, depending on the mission's requirements. By understanding the principles of weight management and aerodynamics, hobbyists can tailor their drones to carry specific payloads without compromising flight performance. This adaptability is essential for surveillance, reconnaissance, and even targeting operations, providing users with versatile tools for modern warfare.

As the use of fixed-wing hobby drones in combat environments becomes more widespread, the question of user experience and flight knowledge becomes increasingly critical. Unlike quad-copters, which are more forgiving and easier to fly, fixed-wing UAVs require a greater understanding of aerodynamics, flight stability, and environmental variables. Their advantages—more extended range, greater endurance, and higher speeds—are only fully realised when piloted by individuals with the knowledge to manage them competently across a range of terrains and weather conditions.

In many modern conflict zones, fixed-wing drones are deployed by irregular forces, volunteers, or non-state actors with limited formal aviation backgrounds. These operators often come from hobbyist communities and lack structured training. As a result, many missions are compromised by pilot error, such as stalling during launch, crashing due to misjudged turns, or losing control in turbulent conditions. The complexity of combat zones—with their unpredictable wind patterns, electromagnetic interference, and the need for precise navigation—further exposes the gap between casual drone experience and the demands of tactical UAV operations.

To mitigate these risks, there is a growing need to transition drone operators from informal hobbyists to trained pilots. A structured approach, including short but intensive instruction from flight trainers or UAV specialists, could dramatically improve mission outcomes. Even a foundational understanding of flight dynamics—such as lift, drag, angle of attack, and centre of gravity—can make the difference between a successful sortie and a lost drone. Real-time decision-making under pressure also improves with formal training, allowing pilots to better manage unexpected gusts, terrain-induced turbulence, and navigation challenges.

Incorporating simulator training, hands-on sessions with an instructor, and scenario-based exercises that mimic battlefield conditions would also increase a pilot's situational awareness. Moreover, it would help build intuitive recognition of flight states—such as distinguishing between a mild pitch oscillation and an imminent stall—and improve reaction time. Recognising the value of such training not only preserves assets but also enhances intelligence-gathering and strike capabilities in contested zones.

In summary, the growing role of hobby kit based fixed-wing UAVs in combat scenarios underscores the importance of transitioning operators from untrained enthusiasts to competent, knowledgeable pilots. Structured training, even if minimal, can bridge the gap between amateur flying and the demanding realities of aerial combat, leading to greater resilience, efficiency, and tactical success on the modern battlefield.

Lastly, once again, the ethical implications of using modified drones in warfare cannot be overlooked. As hobbyists venture into the realm of tactical applications, they must consider the legal regulations and moral responsibilities associated with their innovations. Understanding military regulations is crucial to ensure compliance and to navigate the complexities of drone usage in conflict zones. Engaging in discussions about the ethical use of technology in warfare is vital, as it fosters a responsible approach to the deployment of fixed-wing UAVs and ensures that their capabilities are harnessed for the right purposes.

1.4 Rotary vs Fixed-wing Drone - Tactical Scenarios

In the realm of tactical scenarios, the comparative advantages of rotary and fixed-wing unmanned aerial vehicles (UAVs) are significant, particularly when adapted from hobbyist models. Rotary drones, known for their vertical take-off and landing (VTOL) capabilities, offer unparalleled manoeuvrability in confined spaces and urban settings. This flexibility allows them to conduct reconnaissance missions, provide real-time intelligence, and engage targets from various angles. In contrast, fixed-wing drones excel in endurance and range, making them ideal for extended surveillance operations over vast terrains. The choice between these two types of UAVs often depends on the specific operational requirements and the environment in which they will be deployed.

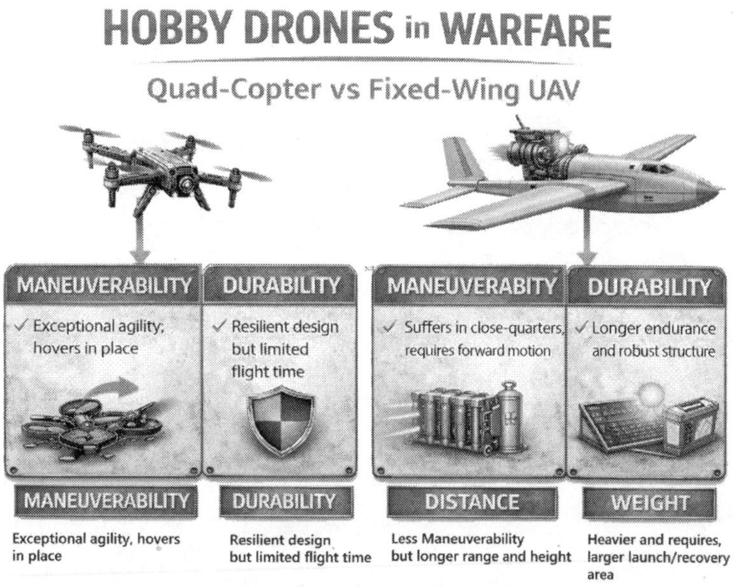

In combat scenarios and active battlefield engagements, the tactical comparison between quad-copters and fixed-wing drones centres on immediate control versus sustained reach. Quad-copters generate lift through continuous rotor thrust, enabling vertical take-off and landing, hover capability, and precise movement in any direction. On the battlefield, this translates into exceptional utility in close-quarters and complex terrain—urban streets, trench systems, wooded areas, shattered infrastructure, and concealed launch sites—where space is restricted and precision positioning is essential. They can rise from cover, hold over a suspected position, descend to avoid detection, and reposition rapidly as the tactical picture evolves. Their ability to remain stationary allows for direct observation of enemy movement, adjustment to shifting threats, and close-in support roles where timing and agility are decisive. In fluid engagements, this manoeuvrability provides a distinct advantage during short-duration reconnaissance and immediate-response situations. The trade-off, however, is endurance. Continuous mechanical lift consumes significant energy, limiting operational range and time on station. In extended engagements, quad-copters must operate relatively close to launch points and are constrained by battery capacity.

Fixed-wing drones, by contrast, operate on aerodynamic lift generated through forward motion across the wings, making them substantially more energy-efficient once airborne. In open battlefield conditions, this efficiency becomes tactically decisive. Fixed-wing platforms excel in long-range reconnaissance, wide-area surveillance, and extended patrol missions, where maintaining persistent coverage over large distances is critical. They can remain airborne for significantly longer periods, travel deeper into contested territory, and monitor supply routes, troop concentrations, or movement corridors well beyond the effective radius of most quad-copters. Their stable cruise profile supports sustained observation and intelligence gathering over distance. However, fixed-wing drones cannot hover and require suitable launch and recovery space or equipment, which can limit flexibility in dense or obstructed environments. During close engagements or rapidly changing ground conditions, they lack the immediate positional control that quad-copters provide.

In battlefield employment, quad-copters dominate in proximity operations where manoeuvrability, concealment, and precision positioning are critical during active engagements. Fixed-wing drones dominate in extended operations where endurance, reach, and sustained surveillance shape the broader operational picture. The relative advantage of each platform is therefore dictated by engagement distance, terrain complexity, and the duration and objective of the mission rather than by inherent superiority of design.

1.5 Marine-Based UAVs

Marine-based Unmanned Aerial Vehicles (UAVs) represent a significant evolution in naval and littoral warfare, especially when integrated with advanced hobby-grade drones—an often underestimated category within modern conflict environments. These adaptable platforms provide expanded surveillance, reconnaissance, targeting, and precision strike capabilities. As global sea lanes and coastal regions become increasingly militarised—from the South China Sea through to emerging Arctic corridors—the operational importance of UAVs deployed from ships, submarines, and offshore installations continues to expand, reshaping the tactical and strategic calculus of modern maritime power projection.

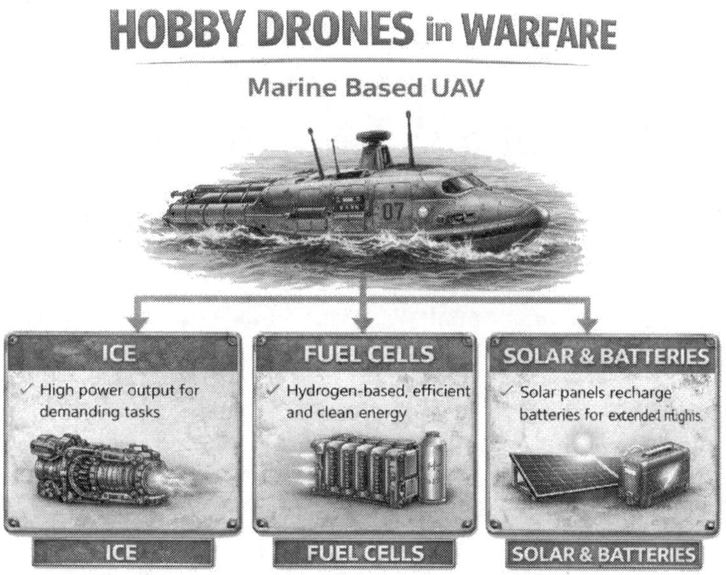

The integration of UAVs with naval platforms extends operational reach, strengthens intelligence-gathering capabilities, and reinforces multi-domain warfare strategies by linking maritime mobility with aerial surveillance. This book explores the development, deployment, and expanding applications of marine-based UAV systems across both hobby-derived and military platforms, examining their role within evolving military operations, including the increasing use of marine-based hobby drones in contemporary conflict environments.

The historical development and use of aerial drones in naval contexts dates back to the World Wars, when early radio-controlled aircraft were experimented with for target practice and reconnaissance. However, the 21st century marked the real beginning of sophisticated UAV integration into naval warfare.

Key Milestones in the Development of Marine UAVs surprisingly dates back to the beginning of the first world war and is often overlooked by many enthusiasts and developers.

Evolution and Application of Marine-Based UAVs, Including Hobby Drone Adaptations

The evolution of marine-based Uncrewed Aerial Vehicles (UAVs)—including both purpose-built systems and adapted hobby drones—has been shaped by key milestones across military, civilian, and technological domains. While modern maritime UAVs are often associated with advanced naval platforms, their development reflects a gradual convergence of military innovation, civilian experimentation, and the increasing accessibility of hobby-grade marine drone technology.

Early interest in aerial reconnaissance over maritime environments dates back to World War I, when tethered balloons and rudimentary radio-controlled aircraft were deployed for naval surveillance. However, these early systems lacked autonomy, endurance, and reliability. True UAV capability in the maritime domain began to emerge only in the late Cold War period, particularly during the 1980s and 1990s, driven primarily by military research and operational necessity.

A major breakthrough occurred in the 1980s with the U.S. Navy's adoption of the RQ-2 Pioneer, a joint U.S.–Israeli development. Launched from naval vessels, the Pioneer provided real-time intelligence, surveillance, and targeting data during the Gulf War in the early 1990s. Its operational success demonstrated the value of UAVs in maritime conflict zones and catalysed global investment in ship-launched aerial systems.

In parallel with military advances, civilian and commercial interest in maritime UAV applications began to grow. Early civilian uses included search and rescue (SAR), fisheries monitoring, and oceanographic data collection. However, these systems remained expensive and limited to institutional users. At this stage, hobby drones were largely confined to recreational flying and short-range photography, lacking the endurance, weather resistance, and navigation reliability used for maritime operations.

By the early 2000s, advances in miniaturisation, GPS integration, and onboard autonomy enabled smaller UAVs to operate effectively from ships and coastal bases. A key milestone during this period was the ScanEagle, developed by Insitu (a Boeing subsidiary). Originally conceived for tuna fishing operations, ScanEagle was rapidly adapted for military intelligence, surveillance, and reconnaissance (ISR). Its long endurance, compact launch-and-recovery system, and ability to operate from frigates and patrol vessels made it a defining example of modern marine UAV design.

The 2010s marked a significant turning point with improvements in battery technology, composite materials, sensor payloads, and flight-control software. These advances directly influenced the hobby drone sector, enabling commercially available multi-rotor and fixed-wing platforms to be adapted for maritime use.

Marine-based hobby drones—often modified with corrosion-resistant coatings, flotation systems, extended antennas, and waterproof housings—began to appear in roles such as coastal surveillance, SAR assistance, environmental monitoring, and disaster response.

At the same time, major naval powers accelerated the development of purpose-built maritime UAVs. Vertical Take-Off and Landing (VTOL) systems such as the Schiebel Camcopter S-100 and Northrop Grumman MQ-8 Fire Scout eliminated the need for runways or catapults, making them suitable for deployment from smaller vessels. These platforms integrated radar, EO/IR sensors, and electronic warfare capabilities, defining a new class of shipborne UAV optimised for harsh marine environments.

In the late 2010s and early 2020s, the convergence of artificial intelligence (AI), machine learning, and autonomous navigation further expanded the role of marine UAVs. Swarming behaviour, vision-based navigation, and GPS-denied operation became increasingly viable. These capabilities are now appearing not only in military systems but also in advanced hobby and prosumer platforms, which can execute semi-autonomous patrols, automated mapping, and object detection over water.

A further milestone has been the emergence of hybrid and multi-domain UAVs capable of operating across air, surface, and sub-surface environments. Systems such as DARPA's Manta Ray and China's "Flying Fish" UAV exemplify this trend, transitioning between flight, flotation, and underwater movement. Although these systems remain primarily military or research-focused, similar concepts are increasingly explored in experimental and hobbyist communities.

Classification of Marine-Based UAVs (Including Hobby Platforms)

Marine-based UAVs—ranging from military-grade systems to modified hobby drones—can be classified according to launch platform, function, and operational capability.

Ship-Launched Fixed-Wing UAVs

These systems provide long-endurance, wide-area maritime surveillance and are typically launched via catapults or rail systems and recovered using nets or arresting gear. The ScanEagle remains a benchmark, offering 20+ hours of endurance and persistent ISR over Exclusive Economic Zones (EEZs). Fixed-wing hobby UAVs adapted for maritime use follow similar principles but operate at reduced scale and endurance.

VTOL UAVs

VTOL platforms are highly valued in maritime environments due to their ability to operate without runways. Systems such as the Camcopter S-100 and MQ-8 Fire Scout are complemented at the lower end by multi-rotor hobby drones deployed from small boats or coastal platforms. These hobby systems are increasingly used for short-range reconnaissance, SAR overwatch, and environmental inspection.

Tethered UAVs

Tethered drones receive power and data connectivity through a physical cable, enabling persistent surveillance limited only by shipboard power supply. While mobility is restricted, these systems—both military and commercial—are highly resilient in contested or GPS-denied environments. Evaluated in conjunction with (FPV) Goggles with fibre optic connections in this book.

Autonomous and AI-Enabled UAVs

Advanced marine UAVs increasingly rely on AI-driven navigation, object recognition, and mission planning. Swarming and coordinated flight are now feasible at both military and hobbyist levels, particularly for mapping, patrol, and disaster response tasks.

Hybrid Aerial-Surface-Subsurface UAVs

Hybrid UAVs can land on water, deploy sensors, or submerge for underwater reconnaissance before returning to flight. These platforms are particularly valuable for anti-submarine warfare, mine detection, and marine science. While still emerging, hybrid concepts are increasingly explored in experimental hobby and research communities.

Shore-Launched Maritime UAVs

Shore-based systems such as the MQ-9B SeaGuardian provide long-range maritime surveillance hundreds of kilometres offshore. At a smaller scale, modified hobby drones launched from coastal facilities play an important role in compliance monitoring, pollution detection, and emergency response.

Enabling Technologies and Strategic Impact

Marine-based UAVs rely on corrosion-resistant materials, hardened air-frames, compact launch-and-recovery systems, and resilient navigation solutions capable of operating in GPS-denied environments. Payloads commonly include EO/IR sensors, radar, AIS receivers, and electronic warfare suites. AI-driven processing enables real-time decision-making and autonomous operation, significantly reducing operator workload and catastrophic failure.

The integration of marine UAVs—ranging from hobby platforms to advanced autonomous systems—has transformed maritime operations by expanding surveillance reach, accelerating response times, and reducing risk to human personnel. Their evolution from simple remotely piloted aircraft to AI-enabled, multi-domain systems reflects the growing dependence of modern militaries, governments, and industries on unmanned technologies.

PETER SAVAGE

Summary

Marine-based UAVs now encompass a broad spectrum, from lightweight hobby drones adapted for coastal and near-shore operations to sophisticated military systems capable of autonomous, long-endurance missions in contested environments. They are defined not only by size or range, but by how they are launched, the missions they perform, and the technologies that enable their operation.

As maritime competition accelerates across the Indo-Pacific, Europe, the Arctic, and other strategically contested regions, marine-based UAVs—whether purpose-built military systems or adapted hobby-derived platforms—are set to remain integral to defence preparedness, environmental monitoring, and maritime security operations. Their adaptability, scalability, and relatively low deployment cost make them particularly valuable in an era defined by persistent surveillance and grey-zone conflict. As these systems continue to evolve in endurance, autonomy, and sensor integration, they will assume an increasingly decisive role in safeguarding sea lanes, projecting maritime power, and protecting critical undersea infrastructure, energy assets, and global communication networks well into the future.

CHAPTER 2

The Rise of Hobby Drones in Modern Warfare

2.1 Hobby Drones in Military Applications

Hobby drones, initially designed for recreational use, have rapidly evolved into valuable assets in military applications. As technology has progressed, these unmanned aerial vehicles (UAVs) have become increasingly capable of fulfilling various roles in combat scenarios. Their accessibility and affordability have made them enticing options for military forces looking to enhance their operational capabilities while keeping costs in check. This overview examines how hobby drones are being repurposed for tactical advantages, focusing on both rotary and fixed-wing designs.

Example -1 Quad-Copter

Quad-Copter modified to carry Ordinance

One of the primary motivations for utilising hobby drones in military settings is their adaptability. Many military units have adopted DIY modifications to enhance the durability of these drones, ensuring they can withstand the rigours of combat environments. Upgrading components such as motors, frames, and batteries allows for extended flight times and increased payload capacities. Additionally, the potential for autonomous navigation systems has opened new avenues for tactical drone operations, enabling these devices to perform complex missions with minimal human intervention, thus reducing risks to personnel and increasing mission success.

Payload optimisation relative to lift capacity remains a critical consideration when adapting quad-copter hobby drones for military application. The weight, balance, and configuration of any payload must be carefully engineered, as even minor changes significantly affect thrust efficiency, endurance, and flight stability. While these UAVs are commonly equipped with cameras and sensors for intelligence, surveillance, and reconnaissance roles, many have been modified to carry ordnance, requiring substantial redesign of propulsion systems, structural reinforcement, and power management to sustain rotary lift and operational range. Careful payload customisation enables operators to configure drones for reconnaissance, target acquisition, electronic warfare, or precision strike tasks. When combined with modular architecture, this flexibility allows rapid adaptation to changing battlefield requirements and improves situational awareness. The integration of artificial intelligence and machine learning further enhances capability, enabling real-time data processing, adaptive navigation, and increasingly autonomous decision-support during missions.

Operational case studies demonstrate how hobby-derived drones have been employed effectively in front-line environments, supporting surveillance, target designation, logistics assessment, and direct combat assistance. These deployments illustrate both the tactical advantages and practical limitations of lightweight UAV platforms under real battlefield conditions. As warfare continues to evolve toward distributed, technology-driven engagements, the operational relevance of hobby drones is expected to expand, reinforcing the need for continued technical refinement as well as serious consideration of the ethical and regulatory implications surrounding their military use.

2.2 Evolution of Hobby Drones

The evolution of hobby drones has undergone a remarkable transformation since their inception, shifting from simple recreational devices to sophisticated tools with significant tactical applications. Initially designed for aerial photography and personal enjoyment, hobby drones have become increasingly valuable in various fields, including front-line military operations. The advancing technology within the (UAV), unmanned aerial vehicle, sector has led to the development of rotary and fixed-wing drones that are not only user-friendly but also capable of complex tasks. This evolution reflects a growing recognition of the potential for hobby drones to enhance tactical capabilities on the battlefield.

From the outset of the Ukraine conflict in 2014, hobbyists began experimenting with do-it-yourself modifications, quickly discovering that durability in combat conditions was paramount. The resilience of small UAVs under fire, in poor weather, and across rough terrain became a central focus. Through hands-on trial and adaptation, users strengthened air-frames, refined motor configurations, upgraded battery systems, and introduced improved materials capable of withstanding vibration, impact, and sustained operational stress. This grassroots innovation culture produced drones increasingly suited to tactical environments, demonstrating that relatively simple platforms could be reinforced to survive the rigours of front-line deployment while retaining agility and efficiency.

Alongside hardware evolution, advances in autonomous navigation emerged as a decisive force multiplier. Enhanced flight controllers, GPS integration, terrain mapping, and sensor fusion enabled hobby drones to execute complex routes with limited human input, significantly improving reconnaissance and intelligence-gathering capabilities. These systems allowed UAVs to manoeuvre through contested or obstructed environments, maintain stable flight under electronic interference, and conduct pre-programmed missions with growing precision. The integration of semi-autonomous and AI-assisted functions marked a substantial shift in battlefield utility, enabling faster data processing, adaptive routing, and near real-time decision support under operational pressure.

At the same time, regulatory and legal considerations struggled to keep pace with the rapid militarisation of modified hobby drones. As these systems became embedded in combat operations, operators faced increasingly complex questions regarding airspace control, privacy, targeting authority, and compliance with domestic and international law. In active war zones, however, legal frameworks often blurred under operational urgency, with immediate survival and mission effectiveness taking precedence over longer-term regulatory scrutiny. The ethical implications of deploying low-cost, easily modified UAVs further complicated this evolving landscape.

PETER SAVAGE

As battlefield demands intensified, payload optimisation became a critical technical priority in transitioning hobby drones from recreational tools to tactical assets. Redesigning commercially available kits for military use became almost instinctive, driving rapid innovation in modular construction and weight distribution. Enhanced lift capacity enabled the carriage of specialised equipment, including surveillance optics, communications relays, and targeting systems, expanding mission versatility. Enthusiasts and military operators alike refined methods to maximise payload efficiency without compromising flight performance, ensuring endurance, stability, and responsiveness remained intact. As these adaptations continue to mature, hobby drones are increasingly integrated into structured military strategies, progressively dissolving the boundary between civilian recreation and modern tactical application.

2.3 Hobby Drones Developed for War

Since the beginning of Russia's full-scale invasion of Ukraine in February 2022, one of the most unexpected and innovative aspects of the war began, the exponential development and adaptation of hobby drones by ordinary citizens to meet military objectives. What began as civilian resistance and volunteering soon transformed into a significant contribution to Ukraine's war-fighting capacity. Hobbyists—many with no formal military background, have played a vital role in reconnaissance, logistics, and even offensive operations using drones that were once the preserve of backyard enthusiasts and YouTube filmmakers.

As Russian forces poured across Ukraine's borders, Ukrainian civilians rose to defend their homeland in every way they could. Alongside volunteers joining the military, another type of resistance emerged—technological. A civilian revolution in the sky began with people experienced in drone flying, model aircraft, and FPV (First Person View) racing began adapting commercial off-the-shelf (COTS) drones such as DJI Phantoms, Mavics, and racing quad-copters into tools of war. The commercial drone industry, which had boomed worldwide for recreational use, was now being weaponised.

The initial application of drones by civilians was reconnaissance. Using consumer-grade drones with built-in cameras, hobbyists provided real-time battlefield intelligence to the Ukrainian military. These drones enabled units to observe enemy movements, adjust artillery fire, locate Russian positions, and assess damage after strikes. In the absence of sophisticated military UAVs at the tactical level, these civilian drones became indispensable.

From toys to tools of war. The most striking evolution, however, has been the transformation of drones into offensive weapons. Hobbyists began modifying quad-copters to carry small payloads—typically grenades or mortar rounds—dropped over enemy positions. FPV drones, originally used for high-speed racing, became kamikaze drones with the addition of explosive warheads and basic guidance systems.

This shift represents a remarkable fusion of civilian creativity and necessity under the pressure of conflict. 3D-printed components, improvised mounts, modified wiring, and repurposed batteries became the norm. Many operators used social media platforms to share designs, techniques, and software configurations, accelerating innovation across front-line units. What would take years in traditional military procurement was happening in weeks—driven by collaboration and urgency.

The war has seen an explosion in FPV drone use, with hobbyist-made suicide drones becoming a cost-effective counter to expensive Russian armour. A $500 USD, FPV drone carrying a shaped charge can disable or destroy a vehicle worth hundreds of thousands of dollars. The asymmetric value exchange has altered battlefield economics, democratising lethality and putting powerful tools in the hands of small units or even individuals.

The use of hobby drones in the Ukrainian war has changed the concept of how conflict is fought forever. What was once a space reserved for highly trained military operators with multi-million-dollar equipment is now accessible to civilians with consumer-grade technology and a laptop. Ukraine's integration of hobbyist drones into combat—guided by volunteers, gamers, engineers, and racers—has not only reshaped battlefield tactics but has fundamentally altered the balance of power in modern conflict. Warfare is no longer limited by expensive arsenals and institutional command. Instead, it now includes grassroots innovation, real-time data sharing, and highly decentralised, low-cost drone swarms. The battlefield of the future is being written not in defence labs, but in basements, workshops, and open-source forums—making it faster, flatter, and far more unpredictable.

The Ukrainian drone effort has not been driven by state procurement alone. Crowdfunding, community, and civilian logistics around the world have raised funds to supply drones, spare parts, batteries, and tools to drone operators on the front lines. NGOs, diaspora communities, and online influencer's have played a critical role in keeping the supply chains active. The United-24 initiative, launched by President Volodymyr Zelenskyy, includes a "Drone Army" program that coordinated public support and centralised distribution of drone technologies.

Moreover, Ukrainian drone schools have been set up by both the military and civilian communities to train thousands of new drone pilots. Some instructors are themselves former school teachers, teenagers, FPV racers or YouTube drone enthusiasts who have stepped into the role of wartime trainers. Their knowledge of aerodynamics, radio signals, and drone control systems has become part of the national defence infrastructure.

Hobby drones, while highly flexible and accessible, come with notable limitations that distinguish them from professional military-grade UAVs. Their range, flight time, and payload capacity are significantly lower, making them less capable in traditional roles of sustained surveillance or heavy-lift operations. However, these limitations have not deterred Ukrainian hobbyists. On the contrary, they have sparked a wave of grassroots innovation, as individuals and small teams across the country worked to push the boundaries of what commercial drones can achieve under combat conditions.

One of the primary challenges has been range. To extend the operational distance of their drones, many hobbyists have replaced stock transmitters and receivers with long-range radio modules. These modifications have enabled drone pilots to maintain control well beyond the original limits—often exceeding 10 to 20 kilometres—allowing safe operation from behind the front lines and deeper penetration into enemy-held territory. In some cases, range endurance was further supported through improvised logistics solutions, including daisy-chained solar charging stations that enabled forward battery replenishment, effectively expanding operational reach and sustaining prolonged deployment cycles in austere environments.

Power systems have also seen extensive refinement. Standard Lithium Polymer (LiPo) batteries, commonly used in racing drones, are optimised for short bursts of high power rather than endurance. Ukrainian drone teams have adapted these systems by experimenting with different battery configurations for longer flight times and creating field-capable setups for rapid battery swaps. These changes have allowed drones to be redeployed quickly with minimal downtime, a critical factor during ongoing combat operations.

Electronic warfare has presented one of the most persistent threats. Russian forces have deployed sophisticated signal jamming and GPS disruption technologies to down or disable Ukrainian drones. In response, drone operators have turned to frequency-hopping spread spectrum systems to avoid jamming, reverted to analogue video feeds where necessary, and integrated autonomous navigation systems using pre-programmed waypoints to ensure mission success even if control links are lost.

Payload adaptation has perhaps seen the most visible transformation. With the help of 3D printers and open-source software, hobbyists have developed custom bomb release mechanisms, lightweight mounting systems, and even stabilised camera platforms for real-time target observation. Some drones now follow pre-set routes or use visual tracking to guide themselves to targets, making them more effective and reducing the pilot's workload.

Together, these technical and tactical innovations demonstrate a remarkable ability to overcome the limitations of hobby drone technology. They reflect a culture of adaptation under fire—one where civilian ingenuity is not just supplementing traditional warfare but reshaping it entirely. Operators have also developed new tactical doctrines. A typical kamikaze FPV strike might involve a spotter drone providing over-watch while the pilot guides the explosive drone to its target. Drone teams often operate in small, mobile units using off-the-shelf goggles, laptop-based planning software, and encrypted messaging apps to coordinate.

While Ukraine's use of drones is largely justified as part of its national self-defence, the precedent it sets is far-reaching. It challenges conventional arms control frameworks and blurs the line between civilian and military technologies.

Operators of modified drones are technically civilians under international law, yet they have engaged in combat operations. This creates ambiguity around their status if captured. Moreover, the widespread availability of drone warfare skills may pose post-war risks if such capabilities are repurposed for crime, terrorism, or civil unrest.

That said, the civilian-led drone movement in Ukraine has shown remarkable discipline. Units tend to follow military command structures, observe target identification protocols, and prioritise enemy combatants and hardware. This distinguishes them from non-state actors in other conflicts who may lack such oversight.

The Ukrainian experience offers profound and lasting lessons for the future of armed conflict. First and foremost, it highlights that "civilian innovation is now a strategic asset". States that can effectively mobilise their civilian technological communities—those with expertise in drones, electronics, coding, and rapid prototyping—will hold a decisive advantage in future conflicts. Ukraine has shown that warfare is no longer the exclusive domain of professional soldiers and defence contractors; it now includes makers, gamer's, and engineers operating from garages and workshops.

Another major takeaway is that "low-cost systems can act as powerful force multipliers". A small group of skilled drone operators equipped with relatively inexpensive equipment can have an out-sized impact on the battlefield. Tasks once reserved for elite military units or high-end UAVs—such as precision strikes, surveillance, and target acquisition—are now within reach of civilians armed with little more than commercial drones and ingenuity.

The war also underscores that "digital literacy is a national security asset". Countries with strong maker cultures, open-source communities, and widespread access to digital tools are far better positioned to innovate under pressure. Ukraine's drone success was not the result of top-down planning but emerged from a digitally fluent society capable of rapid adaptation.

Perhaps most striking is the demonstration that "decentralisation works". Distributed drone teams with local autonomy have proven more agile and responsive than traditional top-down military UAV units. They can innovate quickly, adjust tactics in real time, and exploit battlefield opportunities with remarkable speed. This flexible, horizontal model of drone warfare has reshaped military thinking and will likely be studied for years to come.

Ultimately, Ukraine has rewritten the script on 21st-century warfare. With little more than a smartphone, a soldering iron, and a 3D printer, a teenager in Lviv can now deliver a precision strike against a tank in

Bakhmut. It is a sobering and revolutionary development—one that signals a dramatic shift in how wars will be fought, who will fight them, and what tools they will use.

In conclusion, the adaptation of hobby drones by Ukrainian hobbyists has transformed the battlefield and demonstrated the strategic value of civilian expertise in modern warfare. By turning everyday technology into tools of national defence, Ukraine has redefined asymmetric warfare and highlighted the power of grassroots resistance in the face of aggression. What began as a necessity has become a model—one that other nations will study, replicate, and perhaps fear in future conflicts. As the war continues, the drone war led by civilians and hobbyists may well be seen as one of its most enduring and defining features.

CHAPTER 3

Transforming a Hobby Drone for Warfare

3.1 Modifying a UAV Structure for Warfare

Modifying a UAV for warfare involves a multifaceted approach that encompasses both mechanical enhancements and software upgrades. Hobby drones, originally designed for recreation, can be adapted for tactical operations by improving their structural integrity, payload capacity, and navigation capabilities. The process begins with assessing the drone's frame and materials to ensure it can withstand the rigours of combat environments. Reinforcement techniques, such as using carbon fibre or lightweight aluminium, can significantly enhance durability while maintaining the drone's agility. Furthermore, adding protective casings and shock-absorbing components can help mitigate damage from impacts or adverse weather conditions.

One of the critical aspects of modifying a UAV for military applications is integrating autonomous navigation systems. These systems enable drones to perform complex missions with minimal human intervention, allowing for greater operational efficiency. By utilising GPS and advanced inertial measurement units, hobby drones can be programmed for precise flight paths and automated return-to-home functions. Additionally, incorporating obstacle detection and avoidance technologies can enhance mission success rates in dynamic environments. This autonomy is particularly beneficial for reconnaissance missions, where drones can gather intelligence.

Once again, payload optimisation is another essential consideration for modified drones in warfare scenarios. The ability to carry specialised equipment, such as high-resolution cameras, thermal imaging devices, or even small munitions, greatly expands a drone's utility on the battlefield. Understanding weight distribution and centre of gravity is crucial when customising payloads to ensure stability and manoeuvrability. Moreover, modular designs can facilitate quick changes in payload configurations, allowing operators to adapt their drones to specific mission requirements. This flexibility can be a decisive factor in surveillance operations or precision targeting.

Lastly, the integration of artificial intelligence and machine learning into modified drones offers significant advantages in combat and is covered in this book extensively . These technologies can enhance decision-making processes, allowing drones to analyse data in real-time and respond to threats autonomously. For instance, AI can improve target recognition capabilities, enabling drones to distinguish between civilian and combatant targets more effectively. This advancement not only increases operational efficiency but also raises ethical questions about the extent of autonomy granted to drones in lethal engagements. As hobbyists and military enthusiasts explore these modifications, understanding both the technical and moral complexities of UAV warfare becomes crucial.

3.1.1 Drone Durability, Introduction

The top 10 factors the affecting drone durability in warfare in warfare is influenced by a range of interrelated factors, each playing a crucial role in ensuring reliability and mission success under extreme conditions. One of the most important aspects is structural design and material choice. Drones constructed from durable materials such as carbon fibre, reinforced polymers, or lightweight alloys are more resistant to damage from impacts, weather, and general wear. Modular or crash-resistant frames further enhance survivability on the battlefield, allowing for quick repairs and continued operation after minor damage.

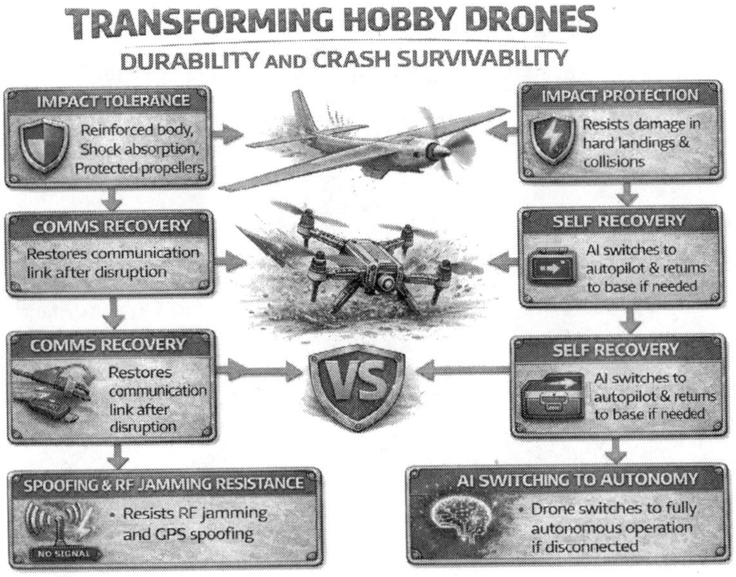

Another critical factor is power system reliability. The durability of a drone is closely tied to its power source. LiPo batteries, while efficient, degrade quickly when subjected to high-stress conditions or extreme temperatures. Internal combustion engines, often used in larger drones, are vulnerable to fuel contamination or mechanical failure. Robust and well-maintained power systems, along with effective power management, can significantly extend a drone's operational life.

Electronic warfare and signal jamming present an increasing threat on modern battlefields. GPS jamming, radio interference, and cyber attacks can disable or misguide drones mid-flight. To endure in contested environments, drones must be equipped with hardened communication links, frequency-hopping technologies, and autonomous fallback systems capable of continuing the mission or returning to base if external control is lost.

Environmental exposure is another major durability factor. Harsh battlefield conditions—such as rain, snow, dust, wind, and temperature extremes—can impair the functionality of electronics, sensors, and motors. Effective weatherproofing, corrosion-resistant components, and thermal insulation are essential to protect the drone and ensure continued performance across varying climates and terrains.

Combat drones also need a high level of impact tolerance and crash survivability. When operating in contested zones, drones may encounter small arms fire, shrapnel, or be forced into hard landings. Designs featuring shock-absorbing frames, foldable or retractable arms, and redundant control systems improve the drone's ability to survive these threats and continue functioning.

Payload integration and balance is another often overlooked factor. Drones that are overloaded or improperly balanced—especially when mounting weapons, cameras, or sensors—can suffer from reduced efficiency, increased motor strain, and unpredictable flight characteristics. Using vibration dampeners and ensuring payload symmetry helps extend air-frame and motor lifespan while improving overall performance.

Maintenance and modularity greatly influence how well a drone can withstand prolonged deployments. Drones built for easy disassembly and repair enable field operators to quickly replace damaged parts, such as motors, propellers, or ESCs (Electronic Speed Controller) , without requiring specialised tools or full rebuilds. Modularity allows for adaptation and upgrading, keeping drones functional even as technology evolves or battlefield needs change.

Closely tied to this is battery health and management. LiPo batteries, while powerful, are sensitive to improper charging practices, excessive use, and harsh environmental conditions. Poor battery handling can lead to swelling, fires, or sudden power loss. Implementing smart charging systems and following disciplined battery rotation routines are essential to preserving battery health and, by extension, the drone's overall reliability.

Software stability and built-in fail-safes also play a major role in drone durability. Stable flight control software with autonomous navigation, return-to-home protocols, and GPS loss handling can prevent crashes due to signal interference or human error. Drones that can operate semi-independently with onboard logic are better suited to survive in high-risk or rapidly changing tactical environments.

Finally, training and pilot skill has become vital to drone longevity. Experienced operators are more capable of navigating complex terrain, reacting to threats, and executing precision manoeuvres under pressure. Poorly trained pilots increase the likelihood of crashes, component failures, and mission losses. Ongoing training, simulation, and field experience directly translate into longer drone service life and more successful deployments.

3.1.4 Power, Longer Missions

Battery technology plays a crucial role in extending the operational capabilities of drones, particularly in combat scenarios where longer flight times can significantly impact mission success. For hobbyists looking to modify their drones for tactical applications, understanding the intricacies of battery enhancements is essential. Lithium polymer (LiPo) batteries, widely used in hobby drones, have seen advancements that allow for higher energy density and improved discharge rates. These enhancements enable longer missions without the need for frequent recharges, allowing operators to gather intelligence or conduct reconnaissance without interruption.

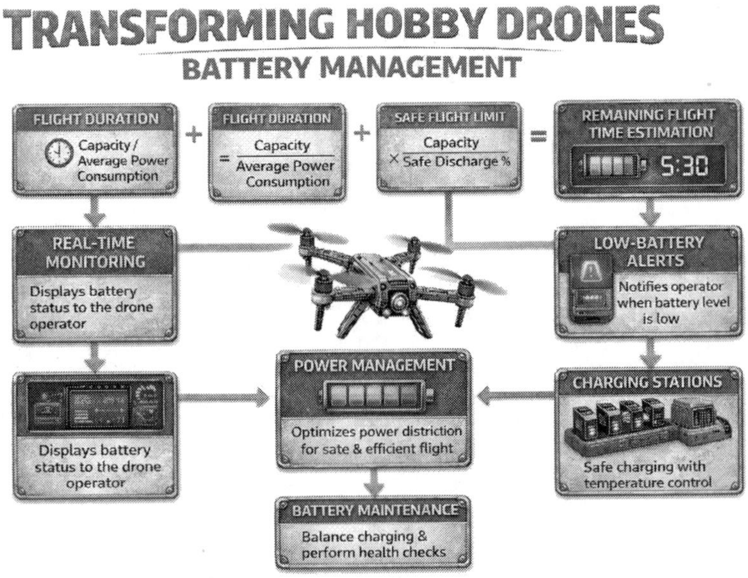

One effective method for achieving longer flight times involves increasing the capacity of the battery while maintaining a manageable weight. Hobbyists can explore high-capacity LiPo batteries that fit within the drone's specifications without exceeding the weight limit. Additionally, implementing parallel connections can enhance overall capacity by distributing the load evenly, thereby reducing the risk of overheating. Careful consideration of the drone's power requirements is necessary, as exceeding the recommended voltage can damage the drone's electronics. By balancing weight and capacity, enthusiasts can maximise their drone's endurance for more extended missions.

Another avenue for enhancement is the incorporation of energy-efficient components. Upgrading motors and propellers to more efficient models can reduce the drone's overall power consumption. Brush-less motors, for example, are known for their high efficiency and longevity, making them ideal for extended missions. Furthermore, optimising the drone's weight by using lightweight materials in DIY modifications can significantly improve flight duration. These adjustments not only enhance battery life but also contribute to improved flight performance, allowing for more agile manoeuvres during tactical operations.

The development of innovative battery management systems (BMS) has also revolutionised how drones manage power. These systems monitor battery health, optimise charging cycles, and provide real-time data on power consumption. Hobbyists can integrate a Battery Management System (BMS) into their drones to ensure efficient use of battery life and prevent over-discharge, which can lead to battery damage. Additionally, some BMS features include regenerative charging capabilities, allowing drones to recapture energy during descent. This innovation can provide a crucial buffer for extended missions, ensuring that drones remain operational longer in the field.

Battery Management System (BMS)

A drone Battery Management System (BMS) acts as the central intelligence for monitoring and protecting the flight battery. It typically comprises three primary functional layers: Data Acquisition (sensors for voltage, current, and temperature), Processing & Algorithms (calculating State of Charge and health), and Protection/ Control (safety disconnects and cell balancing).

- Cell Monitoring (AFE - Analogue Front End): Directly connects to each battery cell to measure individual cell voltages and temperatures.
- Fuel Gauge / MCU: A micro-controller that processes raw sensor data to estimate State of Charge (SoC) and State of Health (SoH).
- Protection Circuitry: Features cutoff MOSFETs (Field-Effect Transistors) to isolate the battery during over-current, over-charge, or thermal events.
- Communication Interface: Typically uses protocols like CAN, UAVCAN, or SMBus to relay battery status to the drone's Flight Management Unit (FMU).
- Cell Balancing: Hardware (often passive resistors) that bleeds excess energy from higher-voltage cells to ensure all cells reach full capacity simultaneously

Lastly, advancements in battery charging technology, such as fast chargers and solar panels, present further opportunities for enhancing drone capabilities. Rapid chargers can significantly reduce downtime between missions, while solar panels provide a sustainable energy source for prolonged operations in the field. This approach not only extends the operational range of drones but also reduces logistical challenges associated with battery replacement. By adopting these battery enhancements, hobbyists can transform their drones into formidable tools for tactical operations, capable of executing longer, more complex missions with greater efficiency.

Together, these ten factors shape how long a drone can survive and operate effectively in combat. Addressing each one through smart design, disciplined operation, and tactical adaptability is essential for maintaining a durable and capable drone fleet in modern warfare.

3.1.2 Strengthening Frame and Components

Strengthening the frame and components of hobby drones is crucial. The unique demands of military applications require modifications that ensure these UAVs can withstand harsh environments and the rigours of tactical missions. Reinforcing frames with lightweight yet robust materials such as carbon fibre or aluminium can significantly increase structural integrity while minimising weight. Additionally, integrating shock-absorbing landing gear can help protect critical components during landings in uneven terrain, ensuring drones remain functional after multiple deployments or recover after a significant impact.

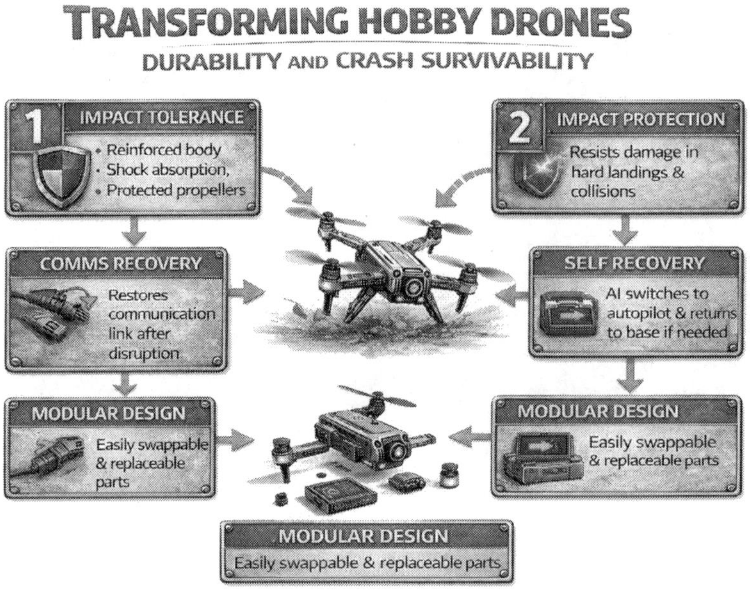

As hobby drones evolve into sustained tactical platforms, the systematic upgrading of core components—frame, motors, batteries, and propellers—has become increasingly essential. Motors designed to withstand prolonged mechanical stress deliver sharper throttle response, improved flight stability, and greater recovery control, all of which are indispensable in fluid combat conditions. Advances in battery technology, particularly the adoption of higher energy-density cells, have extended endurance and expanded operational reach, allowing longer reconnaissance windows without repeated recovery for recharging. At the same time, high-efficiency propellers refine thrust-to-weight performance, enhancing lift capacity and manoeuvrability while preserving energy efficiency. Collectively, these refinements strengthen reliability and survivability in demanding operational environments.

Payload integration, however, introduces a further level of engineering discipline. The careful selection, positioning, and balancing of payload mass are critical to maintaining stable flight dynamics while carrying mission-essential equipment, and these factors must be harmonised with the structural capability of the frame. Although lightweight construction promotes agility and extended endurance, it must be balanced against the need for durability under stress loads and aggressive manoeuvres. Structural resilience, payload weight, and lift capacity must operate in equilibrium to ensure consistent control authority in contested conditions.

While certain components—such as cameras, sensors, and communication systems—may remain relatively consistent, payload configurations can vary significantly between reconnaissance and "hard-kill" missions. Differences in weight distribution and centre-of-gravity placement directly influence flight behaviour, and these dynamics must be calculated prior to deployment. Performance characteristics can shift dramatically the moment ordnance is released; the sudden loss of 600 grams or more alters lift ratios, climb response, and throttle sensitivity, requiring immediate pilot adjustment. Mission flexibility therefore depends on a high-strength yet modular frame design capable of extended deployment and rapid reconfiguration without compromising aerodynamic performance.

Increasingly, the incorporation of artificial intelligence and machine learning into drone architecture is becoming central to this evolution. AI-driven analytics can monitor flight parameters, optimise power management, and adapt control responses in real time, offering valuable decision-support in dynamic tactical environments. As new air-frame designs and flight profiles emerge, intelligent systems will play an expanding role in enhancing operational efficiency, resilience, and mission effectiveness on future battlefields.

3.1.3 Weatherproofing Techniques

Effective waterproofing is fundamental to improving the durability and operational reliability of hobby drones deployed in demanding environments. As these UAVs are increasingly adapted for tactical use, they must endure rain, dust, humidity, and extreme temperature fluctuations without suffering electronic or structural failure. Robust weatherproofing measures significantly extend service life and ensure consistent performance under adverse conditions. By integrating protective treatments and structural safeguards, operators can enhance survivability and reduce mission risk when operating in unpredictable climates.

One of the most widely adopted methods involves applying water-resistant conformal coatings to exposed electronic components. These protective layers shield circuit boards and solder joints from moisture intrusion that could otherwise cause corrosion, signal disruption, or short circuits. All vulnerable electrical connections should be treated prior to deployment, particularly in high-humidity or wet conditions. Complementary sealing strategies—such as silicone sealants or precision-fitted rubber gaskets—can be used to close gaps within the air-frame, preventing the ingress of water and dust while preserving structural integrity.

Protective enclosures for sensitive components represent another critical layer of defence. Custom-built or commercially available weather-resistant housings can safeguard batteries, cameras, GPS modules, and flight controllers from moisture and thermal shock. These enclosures must balance effective sealing with sufficient airflow to prevent overheating, especially during sustained flight operations. At the same time, insulation considerations are essential, as extreme heat or freezing temperatures can significantly degrade battery efficiency and electronic responsiveness. Careful thermal management therefore becomes as important as moisture resistance in maintaining consistent performance.

The incorporation of hydrophobic materials further strengthens environmental resilience. Surface treatments, including water-repellent sprays or advanced hydrophobic coatings, encourage moisture to bead and roll away rather than accumulate on structural or electronic elements. This approach is particularly advantageous in wet or maritime conditions. Selecting corrosion-resistant materials for frames and fasteners also reduces long-term degradation, preserving mechanical strength under prolonged exposure to moisture and dust.

Finally, consistent inspection and preventative maintenance remain indispensable. Regular checks for seal degradation, coating wear, or moisture buildup allow operators to address vulnerabilities early, ensuring that waterproofing systems remain effective and UAV reliability is maintained throughout extended field deployment.

3.1.4 Power, Longer Missions

Battery technology plays a crucial role in extending the operational capabilities of drones, particularly in combat scenarios where longer flight times can significantly impact mission success. For hobbyists looking to modify their drones for tactical applications, understanding the intricacies of battery enhancements is essential. Lithium polymer (LiPo) batteries, widely used in hobby drones, have seen advancements that allow for higher energy density and improved discharge rates. These enhancements enable longer missions without the need for frequent recharges, allowing operators to gather intelligence or conduct reconnaissance without interruption.

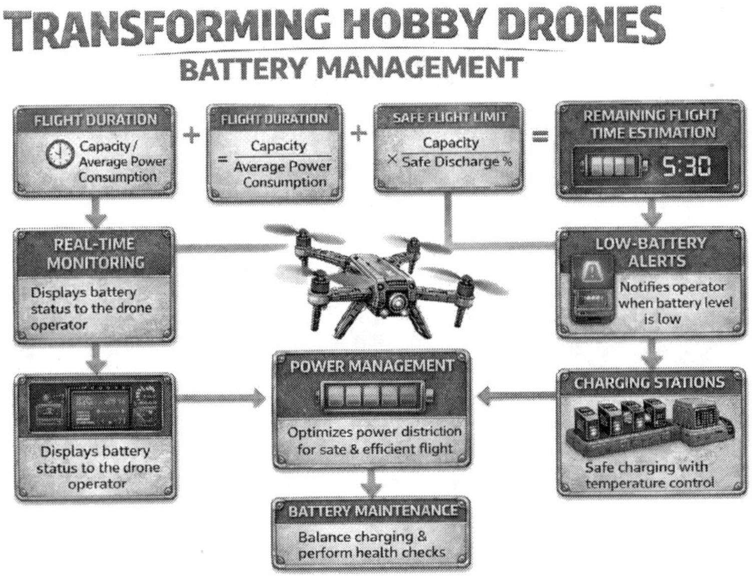

One effective method for achieving longer flight times involves increasing the capacity of the battery while maintaining a manageable weight. Hobbyists can explore high-capacity LiPo batteries that fit within the drone's specifications without exceeding the weight limit. Additionally, implementing parallel connections can enhance overall capacity by distributing the load evenly, thereby reducing the risk of overheating. Careful consideration of the drone's power requirements is necessary, as exceeding the recommended voltage can damage the drone's electronics. By balancing weight and capacity, enthusiasts can maximise their drone's endurance for more extended missions.

Another avenue for enhancement is the incorporation of energy-efficient components. Upgrading motors and propellers to more efficient models can reduce the drone's overall power consumption. Brush-less motors, for example, are known for their high efficiency and longevity, making them ideal for extended missions. Furthermore, optimising the drone's weight by using lightweight materials in DIY modifications can significantly improve flight duration. These adjustments not only enhance battery life but also contribute to improved flight performance, allowing for more agile manoeuvres during tactical operations.

The development of innovative battery management systems (BMS) has also revolutionised how drones manage power. These systems monitor battery health, optimise charging cycles, and provide real-time data on power consumption. Hobbyists can integrate a Battery Management System (BMS) into their drones to ensure efficient use of battery life and prevent over-discharge, which can lead to battery damage. Additionally, some BMS features include regenerative charging capabilities, allowing drones to recapture energy during descent. This innovation can provide a crucial buffer for extended missions, ensuring that drones remain operational longer in the field.

Battery Management System (BMS)

A drone Battery Management System (BMS) acts as the central intelligence for monitoring and protecting the flight battery. It typically comprises three primary functional layers: Data Acquisition (sensors for voltage, current, and temperature), Processing & Algorithms (calculating State of Charge and health), and Protection/ Control (safety disconnects and cell balancing).

- Cell Monitoring (AFE - Analogue Front End): Directly connects to each battery cell to measure individual cell voltages and temperatures.
- Fuel Gauge / MCU: A micro-controller that processes raw sensor data to estimate State of Charge (SoC) and State of Health (SoH).
- Protection Circuitry: Features cutoff MOSFETs (Field-Effect Transistors) to isolate the battery during over-current, over-charge, or thermal events.
- Communication Interface: Typically uses protocols like CAN, UAVCAN, or SMBus to relay battery status to the drone's Flight Management Unit (FMU).
- Cell Balancing: Hardware (often passive resistors) that bleeds excess energy from higher-voltage cells to ensure all cells reach full capacity simultaneously

Lastly, advancements in battery charging technology, such as fast chargers and solar panels, present further opportunities for enhancing drone capabilities. Rapid chargers can significantly reduce downtime between missions, while solar panels provide a sustainable energy source for prolonged operations in the field. This approach not only extends the operational range of drones but also reduces logistical challenges associated with battery replacement. By adopting these battery enhancements, hobbyists can transform their drones into formidable tools for tactical operations, capable of executing longer, more complex missions with greater efficiency.

3.1.5 Battery Life vs Liquid Fuel

The strategic Importance of power systems in drone operations remains highly dependent on the mission type, requirements and desired outcome. The power source in drone operations is far more than a technical detail—it is a strategic factor that directly influences mission success and battlefield effectiveness. For short-range missions, especially in high-threat environments, the need for low noise, quick maintenance, and operational simplicity makes battery-powered drones particularly advantageous. Their silent operation reduces detection risk, and their straightforward logistics support rapid deployment. However, these benefits come with significant trade-offs. Limited energy density and the need for frequent recharging restrict battery-powered drones' endurance, making them vulnerable to mission failure if their power runs out before task completion or return.

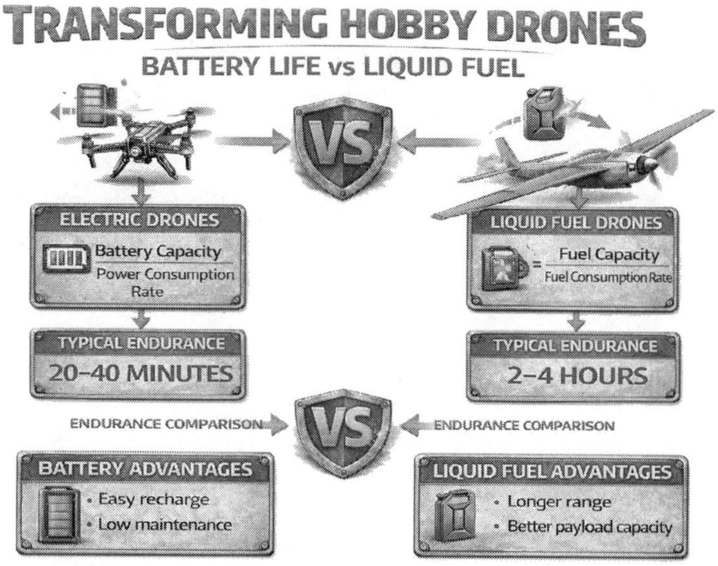

Conversely, internal combustion engine (ICE) or liquid-fuel-powered drones offer extended operational duration and greater payload capacity, but their noise signature increases the risk of detection and engagement, particularly during sensitive operations such as reconnaissance or kamikaze-style missions. A drone emitting the tell-tale sound of an ICE may be identified and neutralised before completing its objective, while one with insufficient battery life may simply never return. These risks underscore how power systems influence not only drone design but also mission planning and tactical employment.

Logistical considerations also play a critical role. Batteries simplify operations in areas where fuel transport is difficult or hazardous, but the infrastructure needed for charging can be limiting in remote or rapidly shifting combat zones. In contrast, while fuel-powered drones offer superior range and flexibility, they require a more complex supply chain and maintenance protocol, factors that may not always be feasible in austere environments.

Battery technology has made substantial advances, especially with lithium-polymer and lithium-ion chemistry, improving energy density and flight times. These improvements have expanded the tactical roles for electric drones, making them well-suited to short-term surveillance, urban operations, or hit-and-run missions. Still, their finite capacity means operational pauses for recharging, potentially interrupting mission tempo and exposing assets to hostile action.

Liquid-fuel-powered drones, on the other hand, are better suited for long-duration missions that demand persistent surveillance, wide-area reconnaissance, or extended loiter times. The use of energy-dense fuels such as gasoline or diesel allows these platforms to remain airborne for hours without the need to refuel or recharge, offering a clear advantage in endurance-focused scenarios. Moreover, their greater payload capacity enables them to carry enhanced sensors, communications gear, or weaponry, further increasing their tactical utility.

In essence, the power system selected for a drone not only defines its technical limits but also dictates its role on the battlefield. Whether prioritising stealth, endurance, logistics, or firepower, commanders must align power source choices with mission objectives, terrain, and threat environment. The strategic importance of these decisions cannot be overstated—they are foundational to achieving operational success in modern drone warfare.

However, integrating liquid fuel systems into hobby drones presents its own set of challenges. The complexity of fuel systems requires additional modifications and maintenance, which can be a barrier for enthusiasts looking to transform their drones for military applications. Moreover, the added weight of fuel tanks and associated components can affect the drone's agility and manoeuvrability, necessitating careful design considerations. Operators need to strike a balance between the benefits of extended flight time and the potential drawbacks of increased complexity and weight.

Ultimately, the decision between battery life and liquid fuel options hinges on the specific mission requirements and operational contexts. For short-range, tactical operations, battery-powered drones may suffice, especially with advancements in energy technology. However, for missions demanding endurance, payload capacity, and extended range, liquid fuel options present a compelling alternative. As the landscape of drone warfare continues to evolve, understanding these dynamics will be crucial for enthusiasts and military operators, enabling maximum effectiveness.

3.1.6 Impact Tolerance , Crash Survivability

Drones operating in contested zones face anti-aircraft fire, shrapnel, or hard landings. Designs that incorporate shock-absorbing mounts, retractable arms, or redundant flight controllers are more likely to recover or continue operating after damage.

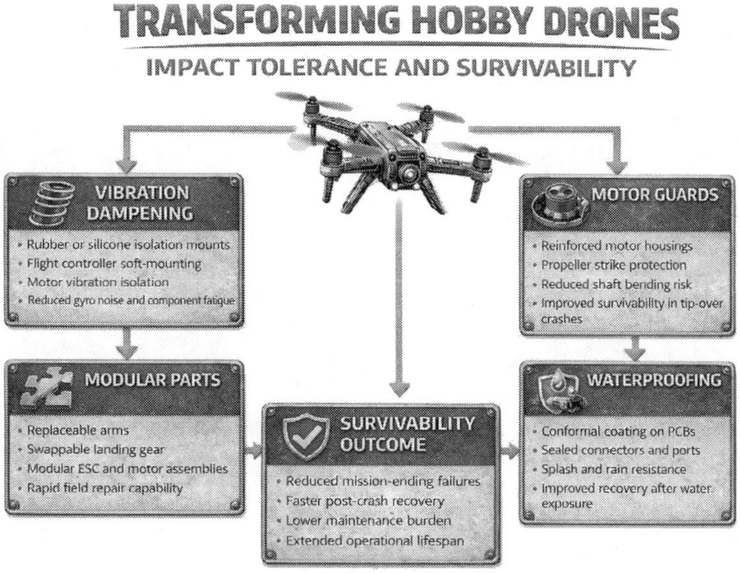

Improving a drone's impact tolerance and crash survivability is essential for extending its operational life, especially in the harsh and unpredictable conditions of warfare. One of the most effective approaches is using durable yet flexible materials like carbon-fibre, reinforced polymers, or TPU, which absorb shocks and reduce structural damage during crashes. Modular design also plays a crucial role—drones built with replaceable arms, detachable components, and isolated electronic compartments are easier to repair and less likely to suffer catastrophic failure. Incorporating shock-absorbing features, such as padded landing gear, vibration-damping mounts for flight controllers, and motor guards, helps cushion hard landings and protect vital systems. Redundant flight systems, including backup motors, ESCs, and sensors, can keep a drone operational even after partial damage. Additionally, programming drones with auto-disarm protocols and emergency landing procedures can limit damage during unexpected impacts. Compact, aerodynamic frames reduce drag and the risk of snagging, while reinforcing high-stress points like motor mounts and battery compartments increases overall toughness. Lastly, pilot training and simulator practice are critical—skilled operators are more likely to recover from erratic flight and avoid crashes altogether. These combined strategies significantly enhance a drone's ability to survive rough handling and return from missions in contested environments.

3.1.7 Payload Integration and Balance

Overloading drones or mounting weapons, cameras, or sensors incorrectly can strain motors and compromise aerodynamics. Well-balanced payloads with vibration dampening reduce wear and increase overall air-frame lifespan.

Improving drone payload integration and balance is critical in warfare, where stability, control, and flight efficiency directly affect mission success. To achieve this, payloads such as cameras, sensors, or munitions must be carefully positioned near the drone's centre of gravity, ensuring that weight is evenly distributed across all axes. Using lightweight materials and custom mounting systems, such as 3D-printed brackets or modular rails, allows for secure attachment without overburdening the air-frame. Incorporating vibration-damping mounts is also essential, particularly for optical or targeting systems, to maintain clarity and accuracy during flight.

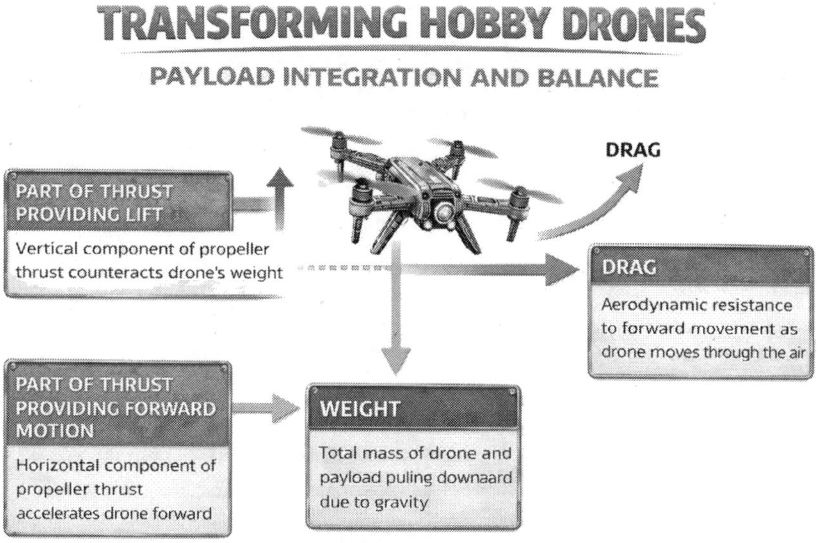

Payloads should be integrated in a way that minimises aerodynamic drag, preserving flight performance and battery efficiency. Additionally, operators should regularly calibrate flight controllers and tune PID settings to account for different payload weights and configurations. In combat conditions, where payloads may need to be swapped quickly, using standardised, quick-release mechanisms can reduce downtime and maintain operational tempo. Thorough pre-flight balance checks and testing in field conditions help prevent instability, motor strain, and mid-mission failures. Overall, a thoughtful and adaptable approach to payload integration not only enhances drone performance but also improves reliability and survivability.

(PID) A drone's PID controller is a feedback loop mechanism using Proportional, Integral, and Derivative control to keep the drone stable and on its intended flight path by continuously adjusting motor speed. The (P) term reacts to the current error, the (I) term corrects for past accumulated errors, and the (D) term anticipates future errors to prevent overshoot and oscillations. Together, these components allow the drone to react to disturbances like wind and maintain its position.

How the PID controller works

Proportional (P): - This term provides a correction based on the current error between the desired and actual position or angle. If the drone is off-target, the (P) term applies a proportional amount of correction, causing an immediate response to level the craft.

Integral (I): This term accounts for the accumulated error over time. It works to eliminate steady-state errors that the (P) term alone can't fix, such as a persistent drift caused by wind.

Derivative (D): This term predicts future errors by looking at the rate of change of the current error. It acts as a damper to prevent the drone from overshooting its target by backing off the correction as it approaches the desired position, which reduces oscillations.

Importance for drones

Stability: PID controllers are essential for maintaining stable flight, allowing the drone to stay level and upright. Response to disturbances: They enable the drone to automatically counteract external forces like wind, preventing it from drifting off course. Precise control: PID control allows for precise maneuvers, holding a steady altitude, or following a complex flight plan utilising a constant performance feedback loop..

Once in flight small disturbances in the environment can affect the drone's altitude. For example, wind gusts may push the drone upward, or the propellers might not deliver exactly the expected thrust. These slight variations can cause the drone to drift a little higher or lower than the desired height.

To correct this, the drone uses intelligent control systems. Three key components of this system are the proportional, derivative, and integral controllers, which will be explained.

Together, these components form the PID (Proportional-Integral-Derivative) control system, which provides precise and stable altitude control. By combining the strengths of each controller, the PID system counters disturbances, corrects errors, and maintains the drone at the desired height. PID control is widely used not only in drone flight systems but in many other engineering and automation applications.

PID In detail

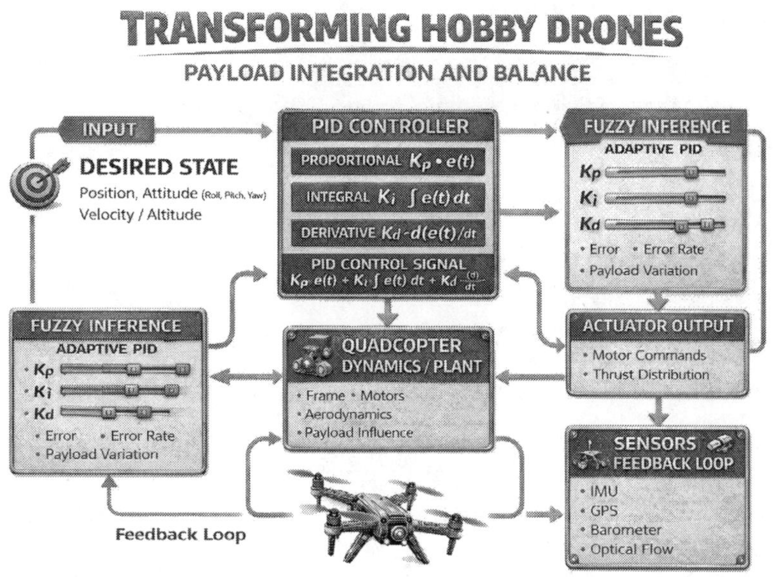

1. Proportional Controller (P-Controller)

A proportional controller provides a corrective action that is directly proportional to the error—the difference between the desired set-point (SP) and the actual process variable (PV).

Using a drone as an example:

You set a desired altitude (SP), and the drone continuously measures its current altitude (PV). If the drone is higher than the set-point, the (P)-controller reduces motor power; if it is too low, it increases motor power. The larger the difference, the stronger the correction.

This makes the drone respond immediately to deviations. However, a (P)-controller alone cannot eliminate all steady-state error, especially in systems affected by persistent disturbances (e.g., wind or payload imbalance).

2. Derivative Controller (D-Controller)

A derivative controller reacts to the rate of change of the error, not just the error itself. It predicts how the system is trending and applies corrective action to counter rapid changes.

In a drone:

Even if the drone is at the correct altitude, it may be rising or falling too quickly. The (D)-controller senses how fast the altitude is changing. If the drone suddenly drops or shoots upward due to turbulence, the (D)-controller applies a corrective response to slow the movement, stabilising the motion before it overshoots the target altitude. This makes the flight smoother, reducing oscillations and helping the drone resist sudden disturbances such as wind gusts.

3. Integral Controller (I-Controller)

Steady-state error is the remaining difference between the desired and actual output after the system has settled over time. A proportional controller can reduce error but not eliminate it entirely. An integral controller solves this by accumulating error over time. The longer and larger the error persists, the more corrective action it applies.

In a drone:

If the drone consistently hovers slightly below the desired altitude due to a small motor or sensor bias, the (I)-controller recognises this persistent error and gradually increases motor power until the height matches the set-point exactly.

Without the I-controller:

- The drone might hover close to—but never exactly at—the desired altitude
- Small, long-term disturbances accumulate and cause steady-state error.
- The drone may oscillate around the desired altitude without ever settling.

With the I-controller:

- Persistent offsets are eliminated.
- The drone can maintain extremely precise altitude, even in the presence of wind or slight mechanical imbalances.
- The control system "learns" from past errors and corrects them.

3.1.8 Maintenance and Modularity

Drones that are easy to disassemble, repair, and upgrade in the field are more durable over time. Modular designs allow quick replacement of arms, motors, ESCs (electronic speed controllers), and other critical parts without requiring full rebuilds.

Improving drone maintenance and modularity in warfare has become essential for ensuring rapid repairs, sustained operations, and long-term reliability in challenging environments. Drones should be designed with modular architecture, allowing key components such as motors, propellers, arms, ESCs, and sensors to be quickly removed and replaced without specialised tools. Using standardised connectors, colour-coded wiring, and labelled parts simplifies field repairs and reduces the risk of error during assembly. Incorporating tool-less or quick-release fasteners can significantly cut down on repair time, especially under combat pressure.

Maintenance efficiency is further enhanced by including diagnostic features, such as onboard error logs, LED status indicators, or companion apps that help operators identify faults quickly. Keeping spare parts kits organised and accessible, along with providing basic training in field servicing, enables teams to keep drones airworthy without relying on rear-echelon support. Additionally, rugged and modular designs help absorb wear and tear while allowing individual parts to be upgraded or replaced as technology evolves. A maintenance-friendly and modular approach ensures that drones can be deployed, recovered, and redeployed efficiently—an essential advantage in the high-tempo operations.

Many experts and drone operators see the adoption of modular components and open-architecture designs in hobby drone warfare as a critical step forward, offering unmatched flexibility and rapid adaptability on the battlefield. Modular systems allow various payloads—such as cameras, munitions, signal jammers, or sensors—to be swapped out within minutes, transforming a reconnaissance drone into a strike platform or an electronic warfare tool with minimal downtime.

This plug-and-play approach significantly reduces the need for multiple specialized drones, streamlining logistics and lowering costs. In fast-paced combat environments, where mission parameters can change rapidly, the ability to reconfigure a drone on-site provides a tactical edge. Whether adapting to terrain, weather, or the type of target, modularity enables forces to respond dynamically without requiring factory-level tools or deep technical expertise. Additionally, damaged drones can be quickly repaired or repurposed using spare modules, increasing operational resilience. Open-source flight controllers and 3D-printed components further support this modular philosophy, allowing irregular or low-budget forces to innovate and deploy mission-specific UAVs in record time. As drone warfare continues to evolve, modular architectures are likely to become a standard feature, bridging the gap between low-cost hobby platforms and high-end military systems through adaptability, speed, and mission versatility.

3.1.9 Battery Health and Management

Drone battery health is one of the most critical and misunderstood aspects of unmanned aerial vehicle performance, reliability, and operational safety. Whether in hobbyist environments, commercial operations, or military and conflict-zone deployments, battery health determines not only how long a drone can stay in the air, but also how safely it can execute its missions. The concept of "battery health" blends several measurable indicators—chemical condition, ability to hold charge, internal resistance, heat behaviour, structural integrity, discharge characteristics, and consistency of cell balancing—into a single assessment of whether a battery can deliver power safely and reliably. Determining and measuring battery health for drone operations therefore requires a multi-dimensional evaluation, combining physical inspection, electronic measurement, cycle-life analysis, and performance monitoring under load. As the UAV industry has evolved, particularly with the accelerated demands of military and dual-use drone technology, battery health assessment has become increasingly sophisticated, with onboard telemetry, AI-driven analytics, and predictive modelling playing an expanding role. Nevertheless, the foundation remains rooted in a scientific understanding of how Lithium Polymer (LiPo) and Lithium-ion (Li-ion) cells age, degrade, and fail.

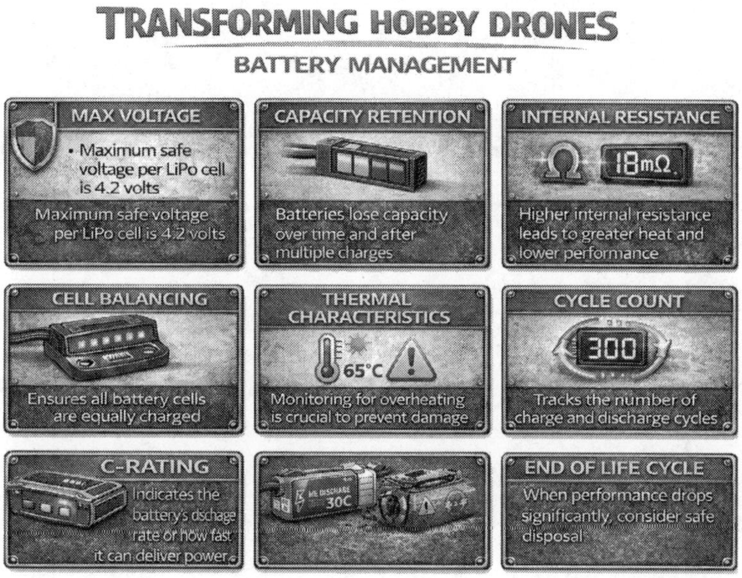

Maximum voltage level

Battery health is first and foremost determined by examining a battery's ability to maintain its nominal and maximum voltage levels. All lithium-based drone batteries are composed of individual cells—typically 3S, 4S, 6S, or 12S configurations—with each cell operating around 4.2V fully charged and approximately 3.7V nominal. When new, a healthy cell reaches 4.20V ± 0.03 during charging and stabilises without drifting downward under resting load. As a battery ages, maximum cell voltage may begin to fall, a phenomenon called voltage sag or capacity fade. If a fully charged cell consistently peaks only at 4.15V or 4.10V after a complete balance charge, it strongly indicates chemical degradation. Measuring battery voltage consistency across all cells during and after charging is one of the simplest and earliest methods of determining battery health. High-quality chargers and UAV battery management systems record peak cell voltages, monitoring how these change over dozens or hundreds of cycles. A cell that drifts 0.03–0.05V lower than its neighbours is considered slightly aged, while 0.1V or more drift suggests the cell is entering a failure trajectory. Drone operators—particularly in military environments—monitor these numbers closely because voltage drift correlates strongly with unpredictable flight times and mid-air power loss.

Capacity retention

The second fundamental measure of battery health is capacity retention, meaning how much charge the battery can still store relative to its original specification. A 5,000mAh (5Ah) LiPo battery when new typically delivers between 96–102% of its advertised capacity. Over time, electrodes wear, electrolyte additives degrade, and the solid-electrolyte inter-phase (SEI) layer thickens, reducing ion mobility. The battery begins to lose usable capacity, sometimes slowly, sometimes suddenly depending on how it is used, charged, and stored. Capacity is measured during a controlled discharge cycle using a charger or analyser capable of placing a constant load on the battery, typically at 1C (a current equal to the battery's rated capacity—e.g., 5A for a 5,000mAh pack). The analyser discharges the battery from full charge to a safe minimum cutoff, usually 3.0V or 3.3V per cell, measuring how much energy it was able to extract. A healthy LiPo retains at least 90% of its rated capacity for the first 50–100 cycles, depending on how aggressively it is used. As the number drops—85%, 80%, 75%—the, cell balancing, battery becomes progressively less reliable. In military drones, a battery falling below 80% capacity might be relegated to training use due to the increased risk of voltage sag under heavy load. In FPV drones used for reconnaissance or kamikaze strikes, where rapid acceleration and high burst currents are common, even an 85% capacity battery can experience catastrophic collapse mid-flight. Capacity measurement therefore functions as a quantitative benchmark for deciding whether a battery remains mission-ready.

Internal resistance

A closely related but distinct metric is internal resistance (IR), which measures how much the battery resists the flow of current. Internal resistance increases naturally as the battery ages, but it can also spike abruptly when the battery is damaged, over-discharged, overheated, or manufactured with poor quality control. IR is typically measured in milliohms (mΩ) and can be tested at the cell level or the pack level using either built-in charger measurement functions or specialised ESR (Equivalent Series Resistance) meters. New LiPo cells often have IR readings between 1–4 mΩ depending on size and chemistry. As a battery degrades, the internal resistance rises, causing more heat generation under load and reducing the battery's ability to deliver high currents. When IR readings increase by more than 50% from their original baseline, the battery is considered significantly aged. If a cell reads twice the IR of its neighbouring cells, the pack becomes dangerous, as this cell will heat unevenly and may experience thermal runaway under high discharge conditions. Measuring internal resistance is one of the most reliable methods of predicting imminent battery failure because IR changes often appear before capacity loss becomes obvious. In professional UAV systems, onboard telemetry systems continuously monitor IR trends, alerting operators when thresholds are exceeded.

Voltage sag under load

Another crucial measure of battery health is voltage sag under load, which refers to the drop in voltage that occurs when the battery supplies high current during flight. This is one of the most realistic and practical indicators of a battery's operational health. Even if a battery shows acceptable voltage and capacity levels under static testing, a degraded battery will typically sag severely when placed under real flight load, such as during takeoff, full throttle climbs, or sudden manoeuvres. Voltage sag is caused by increased internal resistance, reduced conductivity of electrode materials, and ion transport limitations that worsen as the battery ages. Measuring voltage sag requires either flight telemetry or bench testing with a programmable electronic load. A healthy battery maintains voltage above 3.7V per cell during heavy load. Moderate sag to 3.5V may be acceptable, but dips below 3.3V under normal flight loads indicate serious degradation. For long-range drones, sagging to 3.3V or below can cause the flight controller to trigger emergency landing or power cutoff. For combat drones, severe sag can cause catastrophic failure before reaching the target. Therefore, voltage-sag profiling has become a standard component of drone fleet management, especially for swarming or autonomous systems where large batches of batteries must be evaluated consistently.

Cell balancing

Cell balancing performance is another major indicator of drone battery health. In multi-cell packs, a balance connector allows chargers to equalise the voltage of each cell. When new, cells remain tightly balanced through multiple charge/discharge cycles. As the battery ages or experiences physical or thermal damage, cells begin to drift apart. Some cells charge faster, some slower, and some never reach the same voltage as their neighbours. Poor cell balance indicates uneven internal resistance, inconsistent capacity distribution, or internal cell damage. During charging, a healthy battery requires only minor balancing, but a degraded battery might require extensive balancing—sometimes 20 minutes or more—as the charger struggles to equalise the cells. Persistent imbalance indicates that at least one cell has fundamentally deteriorated and the battery is no longer reliable. In drone warfare applications, any pack showing chronic imbalance is typically removed from operational use because imbalanced cells dramatically increase the risk of thermal runaway or sudden in-flight power loss. Cell balancing behaviour therefore serves as both a diagnostic metric and a validation test for whether a battery should be cleared for continued use.

Physical inspection

Physical inspection also plays a critical role in determining battery health. LiPo batteries in particular are highly susceptible to physical damage, puncture, denting, or swelling (commonly called "puffing"). Puffing occurs when electrolyte breaks down and releases gas inside the pouch cell, causing visible swelling.

Swelling is an unequivocal sign of chemical breakdown and internal pressure increase, both of which make the battery extremely dangerous. A swollen LiPo can ignite or explode under load or during charging. Even small amounts of swelling indicate that the battery should be retired. Physical inspection also includes checking for damaged wires, cracked solder joints, corroded balance leads, insulation failure, or external contamination from water, mud, or fuel—especially relevant in conflict zone operations where environmental exposure is high. Impact damage from crashes is particularly significant: even if the battery appears normal, any hard crash can rupture separators inside the cell, creating a latent short circuit that may only become apparent during the next charge. Therefore, assessing battery health must include a rigorous visual and tactile inspection, combined with a review of the battery's operational history.

Thermal behaviour

Closely related to physical inspection is the evaluation of thermal behaviour, as heat is both a cause and a sign of battery degradation. A healthy battery remains relatively cool during discharge, with moderate warming at high loads. As internal resistance rises with age, more electrical energy is converted to heat rather than usable power. Operators measure battery temperature using infrared thermometers, onboard telemetry sensors, or thermal cameras. A battery that consistently operates at temperatures above 60°C under moderate load is considered unsafe. Additionally, temperature spikes during charging indicate internal shorting, SEI instability, or advanced degradation. Thermal health assessment is especially important in autonomous and long-range drones where onboard thermal telemetry is used to automatically limit flight performance when temperatures exceed safe thresholds. In military systems, AI-enhanced battery management software analyses temperature patterns for early signs of cell failure, allowing predictive replacement before a critical mission.

Drone operators also analyse charge behaviour, including whether a battery accepts and retains charge normally, how long it takes to complete a full charge cycle, and how quickly voltage stabilises after charging. A healthy battery charges consistently in line with its rated capacity and C-rating. Deviations—such as rapidly reaching 4.2V but then losing charge immediately, or taking unusually long to reach full voltage—indicate problems with electrolyte composition or SEI layer growth. Charge acceptance, meaning how efficiently ions intercalate into the electrode material during charging, decreases with age. When charge acceptance falls substantially, the battery may show premature "full charge" indicators despite holding significantly less usable energy. Chargers equipped with advanced analytics can identify irregular charge curves and alert operators to degradation.

Cycle count

Cycle count is another conventional metric used to determine battery health, but on its own it is imprecise. Lithium batteries typically have a finite cycle life—100 to 300 cycles for high-performance LiPos, and 300 to 600 cycles for high-quality Li-ion cells. However, batteries degrade not merely due to cycle count but due to how the cycles are used. High-current draw, deep discharges, frequent storage at full charge, and high temperature exposure all dramatically accelerate ageing. Therefore, while cycle count provides a general sense of where a battery is in its lifecycle, it must be interpreted alongside other measurements such as capacity retention and internal resistance. Some batteries that have been carefully managed can remain healthy well beyond 200 cycles, while others used aggressively in FPV or military drones might deteriorate after fewer than 30 cycles.

Coulomb counting

Another sophisticated metric increasingly used in professional and military drone contexts is coulomb counting, which measures the exact amount of energy entering and leaving the battery across all charge and discharge cycles. This method provides a highly accurate representation of battery efficiency and capacity fade. Coulomb counting systems integrate with the drone's onboard electronics and record current and voltage in real time. Over months of operation, this data builds a comprehensive profile of long-term degradation. For example, a battery that once delivered 80 Wh (watt-hours) but now only delivers 64 Wh has lost 20% of its energy capacity. Coulomb counting is more accurate than simple mAh readings because it accounts for voltage variation and discharge curves. In swarm drones or fleet-scale deployments, coulomb-counting analytics allow operators to track hundreds of batteries simultaneously and predict which ones will fail first.

Performance-based metric

Beyond direct measurement tools, battery health is assessed using performance-based metrics, meaning how the battery behaves during actual missions. Flight time is the simplest and most intuitive indicator. If a drone that previously remained airborne for 25 minutes suddenly achieves only 19 minutes, and all other variables remain constant, the battery is losing capacity. Maximum thrust performance is also affected because power output depends on voltage stability. A worn battery may cause sluggish acceleration, decreased lift, or inability to maintain altitude in wind. Pilots can feel these changes long before numerical tests confirm the degradation. In high-demand environments—such as drone racing, long-range reconnaissance flights, or FPV strike drones used in modern conflicts—operators rely on performance cues to decide when to retire batteries. Rapid deterioration of flight performance is often more decisive than laboratory metrics in real-world decision-making.

Storage health

Furthermore, storage health contributes significantly to long-term battery condition. Lithium batteries degrade more quickly when stored fully charged or fully discharged. Over months, high-voltage storage accelerates electrolyte breakdown, while low-voltage storage risks cell failure or permanent capacity loss. Therefore, monitoring the voltage level at which batteries are stored is essential. Most manufacturers recommend storing LiPo batteries at around 3.80V per cell. Chargers with storage-mode functions use this target voltage to prepare batteries for long-term storage. Deviations from recommended storage practices contribute directly to ageing and distort battery health measurements by creating inconsistencies between tested and actual performance.

Mechanical integrity

In addition to chemical and electrical metrics, battery health assessment increasingly considers mechanical integrity and environmental exposure. Drones used in harsh conditions—dust, water, salt, extreme cold or heat—face accelerated degradation. Moisture ingress can corrode internal tabs or damage protective circuitry. High altitude operations reduce air density, lowering cooling efficiency and increasing battery temperatures. Cold weather thickens electrolytes and temporarily reduces capacity. These environmental stressors affect the interpretation of battery health metrics by exaggerating sag, altering internal resistance, and skewing capacity readings. Comprehensive health analysis therefore contextualises data within environmental histories, allowing operators to distinguish between temporary performance dips and permanent degradation.

C-rating

Another important consideration is the battery's discharge C-rating—a number that indicates how quickly it can safely discharge energy relative to its capacity. High-C batteries can deliver higher current but typically degrade faster due to intense ion movement. C-rating decline is a subtle marker of aging: a battery originally rated at 75C may only perform effectively at 40C after significant use. Measuring C-rating degradation involves monitoring maximum current output, thermal rise, and voltage sag during extreme load tests. In military drones that may require sudden power bursts—such as for rapid ascent to avoid ground fire or to deliver a payload—reduced C-rating can severely compromise mission success. Therefore, operators track C-rating performance as an essential part of battery health analysis.

Impedance spectroscopy

Battery health is also measured through impedance spectroscopy, a laboratory-grade technique increasingly adopted by high-end drone manufacturers. This method applies alternating current signals at various frequencies to measure how the battery responds. The resulting frequency-response curves provide insights into ion transport behaviour, electrolyte condition, SEI stability, and electrode degradation. While not commonly used by hobbyists or small drone units, impedance spectroscopy is used in designing smart batteries for autonomous drones where high reliability is mandatory. AI models trained on spectroscopy data can predict capacity fade months in advance, helping logistics teams plan battery replacements before failures occur.

Battery Management Systems (BMS)

Modern drone ecosystems incorporate smart Battery Management Systems (BMS) that constantly evaluate battery health using sensors and embedded algorithms. These systems track voltage, temperature, current, internal resistance, charge cycles, and other parameters. A smart BMS can shut down charging if unsafe conditions are detected, balance cells automatically, limit current output during critical missions, and log health data for later review. Drone fleets using smart BMS technology enjoy significantly higher reliability, as most catastrophes arise not from sudden failures but from gradual degradation that goes unnoticed. Smart BMS-driven health assessments often include state-of-charge (SOC), state-of-health (SOH), state-of-power (SOP), and residual capacity estimates, all of which contribute to a comprehensive health score.

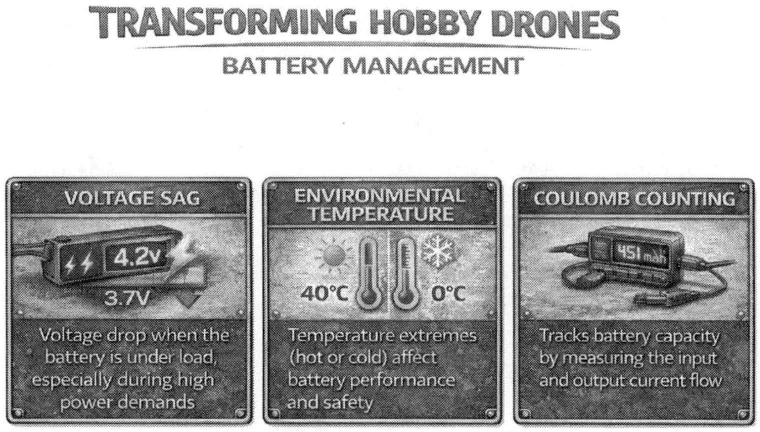

Batteries in conflict zones

In conflict zones, where supply chains are strained and drones are deployed aggressively, battery health assessment becomes even more crucial. Drones operating with degraded batteries risk mission abort, crash landings, or self-destruction short of target. Operators therefore use rapid health triage methods—quick voltage checks, IR spot tests, thermal scans, and flight-performance sampling—to determine whether a battery can be reused. High-usage batteries are often marked, logged, and assigned to progressively less demanding missions as they age. The chaotic nature of battlefield drone usage makes structured battery health measurement essential to maintaining operational tempo and preventing avoidable losses.

Battery health determination also involves understanding failure modes, such as dendrite formation, gas generation, electrolyte breakdown, and internal short circuits. Each failure mode manifests in measurable indicators such as abnormal heat, sudden drops in capacity, or oscillating cell voltages. Predicting these modes requires monitoring trends rather than single measurements. For example, gradually rising internal resistance over several weeks indicates predictable ageing, but sudden jumps indicate serious internal damage. Distinguishing between these patterns helps prevent catastrophic failures.

End of Life Cycle

Because drone batteries are consumables, health measurement also includes end-of-life determination—a formal decision process for retiring a battery. Most operators retire batteries when they fall below 80% of their original capacity, but some retire at 85% depending on mission risk. Extreme sag, swelling, severe imbalance, or temperature instability accelerates retirement. Batteries deemed unsafe for flight may still be used for ground testing or power supply work, but they should never return to airborne operations.

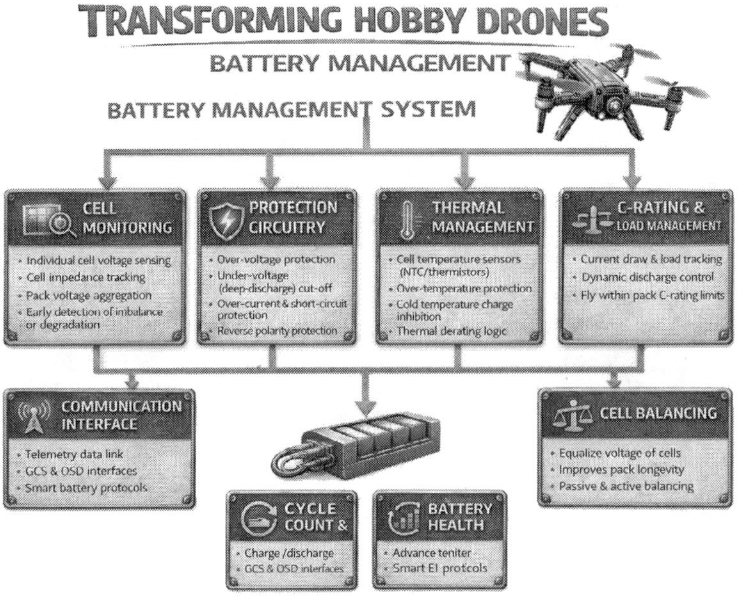

Summary

In summary, drone battery health is determined and measured through a combination of voltage assessment, capacity retention testing, internal resistance measurement, voltage sag behaviour, cell-balance consistency, thermal stability, charge/discharge pattern analysis, coulomb counting, performance-based observations, environmental condition monitoring, mechanical inspection, C-rating retention, and advanced laboratory techniques such as impedance spectroscopy. As drones become increasingly central in civilian, commercial, and military applications, accurate battery health assessment becomes indispensable. Healthy batteries ensure predictable performance, mission success, and operator safety, while degraded batteries pose unacceptable risk. Comprehensive, data-driven battery health measurement practices therefore form the backbone of safe, reliable, and effective drone operations.

3.1.10 Software Stability and Fail-Safes

Reliable firmware and flight software with built-in fail-safes—such as return-to-home, GPS loss handling, or manual override—can prevent mission-ending crashes. Drones with autonomous navigation capabilities also reduce operator error and improve flight corrections.

Improving drone software stability and fail-safes in warfare is essential to ensure reliable, autonomous operation in high-risk environments where human intervention may be limited or compromised. In combat conditions, where drones are exposed to signal interference, GPS denial, and unpredictable threats, software must not only function without failure but also be adaptive, resilient, and capable of autonomous decision-making. One of the foundational steps in achieving this is developing and maintaining robust, well-tested flight control software. This includes continuously updated firmware that can handle variable payload weights, compensate for environmental conditions, and recover from temporary sensor malfunctions or communication loss.

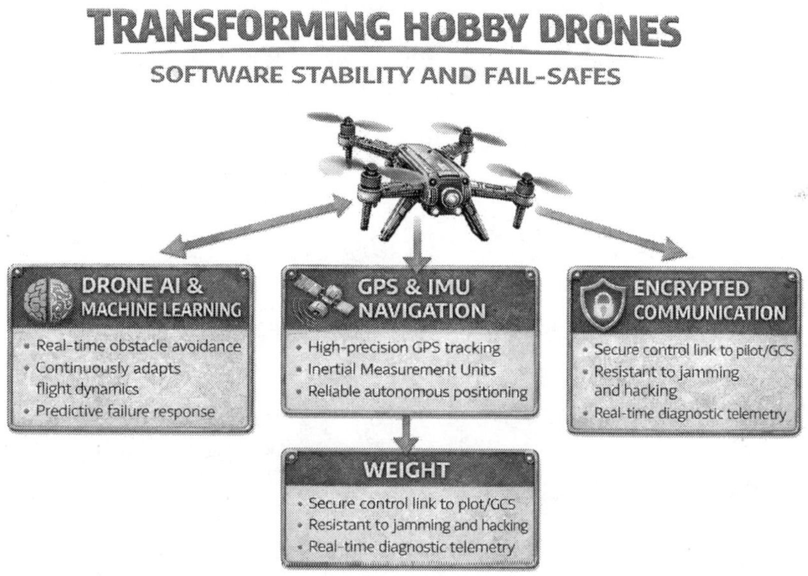

Software should be rigorously tested in simulated combat environments using digital twins or battlefield emulators to expose bugs and vulnerabilities before deployment. Incorporating multiple layers of fail-safes, such as auto-return-to-home, geo-fencing, low-battery landings, and GPS-loss fallback modes, ensures that drones can respond intelligently to emergencies rather than crashing or being lost.

The use of (AI) and machine learning (ML) further enhances software stability and resilience. By training drones on large data sets gathered from previous missions, simulations, and sensor data, AI algorithms can learn to recognise patterns—such as approaching threats, terrain changes, or signal interference—and adapt flight paths or behaviours accordingly. For example, computer vision systems powered by ML can enable drones to navigate visually in GPS-denied environments, identify enemy assets or safe landing zones, and avoid obstacles in real time. Reinforcement learning models can also be applied to train drones in dynamic decision-making, such as choosing the most efficient evasive manoeuvres when under fire or selecting alternate routes in response to detected jamming. These adaptive capabilities make drones far more autonomous and reduce the dependency on constant operator input, which is often disrupted during combat.

To further improve reliability, redundant software systems should be implemented across critical subsystems, allowing drones to fall back on secondary control modes if the primary system fails. This could include switching from GPS-based navigation to inertial measurement units (IMUs) or visual odometry if satellite signals are lost or spoofed. In addition, AI-driven diagnostics can continuously monitor system health during flight, detecting early signs of failure in motors, sensors, or power supply and triggering automatic responses like return-to-base, hover-and-hold, or safe emergency landing.

Software security is another critical aspect of wartime drone operations. Incorporating encrypted communication protocols, firmware authentication checks, and intrusion detection systems helps protect drones from being hijacked or fed false commands. AI can play a role here too, by analysing network behaviour and detecting anomalies that might suggest a cyber attack or spoofing attempt. Furthermore, cloud-based AI platforms—when connectivity allows—can enable real-time coordination of drone swarms, collaborative target tracking, and shared situational awareness, dramatically improving battlefield efficiency.

Ultimately, combining robust coding practices, adaptive AI systems, and layered fail-safes enables combat drones to operate with greater autonomy, accuracy, and survivability. In warfare, where split-second decisions and unpredictable threats are the norm, stable and intelligent software is not just a technical advantage—it is a mission-critical necessity.

3.1.11 Training and Pilot Skill

Operator experience has a direct impact on drone longevity. Skilled pilots are more likely to avoid obstacles, adapt to interference, and recover from malfunctions. Poor piloting, on the other hand, leads to avoidable crashes and excessive wear.

Training and pilot skill are among the most critical factors influencing the effectiveness and survivability of drones in warfare. A well-trained drone operator can navigate complex terrain, adapt to rapidly changing battlefield conditions, and execute precision tasks under pressure, while an unskilled pilot may cause mission failure through avoidable crashes or misjudged manoeuvres. In modern conflict zones, especially in Ukraine and similar theatres, drones are used for reconnaissance, targeting, strike missions, and psychological operations—tasks that require both technical proficiency and tactical awareness.

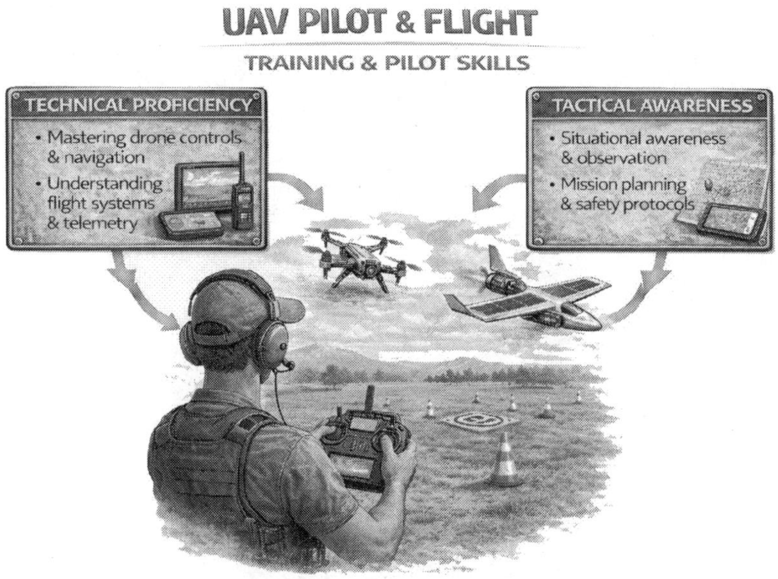

To improve pilot skill, comprehensive and continuous training programs must be established, incorporating both basic flight operations and advanced combat scenarios. This includes simulated training using FPV (First-Person View) simulators, which allow pilots to practice in virtual environments that mimic real-world conditions, including GPS jamming, hostile engagement, and low-visibility situations. Simulators help operators build muscle memory and confidence, enabling faster, more accurate responses in live deployments.

Live field exercises are equally important, providing experience with various drone platforms, payload configurations, and mission types. Operators should be trained not only in piloting but also in emergency recovery procedures, rapid battery swaps, maintenance routines, and basic repairs—skills that increase drone uptime and reduce reliance on rear logistics. Advanced training can also include tactical flight strategies, such as flying low to avoid radar, using terrain masking, or coordinating with infantry units. Moreover, operators need to be familiar with electronic warfare countermeasures, including frequency-hopping techniques, analogue fallback systems, and visual navigation alternatives when GPS is denied.

The growing integration of autonomous and AI-assisted drones doesn't eliminate the need for pilot skill—in fact, it raises the bar. Pilots must now understand how to supervise semi-autonomous systems, program autonomous flight paths, interpret telemetry and video feeds for targeting, and override systems when AI encounters edge cases. As drone warfare becomes more technical, training must also include data analysis, mapping software, and mission planning tools, ensuring that pilots become multi-disciplinary operators, not just remote drivers. Regular after-action reviews and performance analytics can identify weaknesses, reinforce best practices, and allow for rapid skill development. In the high-stakes context of warfare, where drones are force multipliers and frontline assets, investing in operator training is not optional—it is a strategic necessity that directly impacts mission success and battlefield superiority.

CHAPTER 4

Drone Battery Technologies (Li-Po)

4.1 Drone Batteries, Introduction

Batteries remain one of the most critical elements in UAV operation—an issue examined in depth throughout the book—as they provide the primary power source for propulsion, onboard electronics, sensors, and communication systems.. The rapid development of drone technology has gone hand in hand with advances in battery chemistry, energy density, and power management, making modern drones more efficient, longer-lasting, and capable of carrying heavier payloads. Lithium-based batteries, particularly Lithium Polymer (LiPo) and Lithium-ion (Li-ion), dominate the drone market due to their high energy density, lightweight construction, and ability to deliver the high current bursts needed for propulsion and manoeuvring. These batteries provide the essential balance between weight and capacity, directly influencing flight time, range, and overall drone performance. However, despite their advantages, these batteries also present challenges, such as thermal management issues, sensitivity to overcharging or deep discharging, and potential safety risks like fire or explosion if damaged or improperly handled.

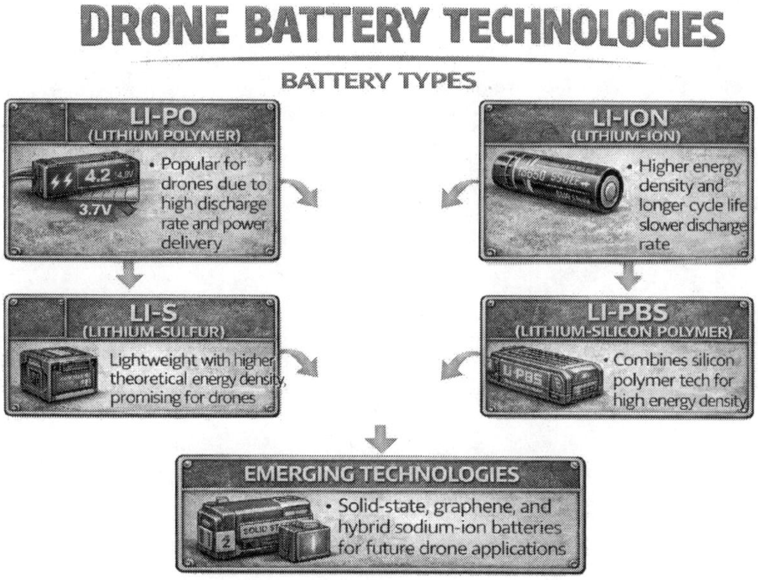

The evolution of drone batteries is also driven by the demands of different applications—from small hobbyist quad-copters to large military or industrial drones. For example, consumer drones typically rely on compact, rechargeable LiPo packs that can provide 15 to 30 minutes of flight time, while more sophisticated commercial or military drones might use advanced battery technologies combined with hybrid power systems to extend endurance significantly. Additionally, emerging battery innovations, such as solid-state batteries and lithium-sulfur chemistry, hold promise for further improving energy density, safety, and charging speed. Alongside chemistry improvements, advances in battery management systems (BMS) play a vital role in monitoring battery health, optimising charge cycles, and preventing damage, thereby enhancing both safety and longevity.

Furthermore, the infrastructure supporting drone battery recharging and replacement is an important consideration, especially for applications requiring extended operations or rapid turnaround. Portable charging stations, battery swapping systems, and solar recharging integration's are all being explored to address these operational needs. The life-cycle and environmental impact of drone batteries also raise important sustainability questions, as the growing drone market increases demand for raw materials like lithium and cobalt. Recycling efforts and the development of eco-friendly battery materials are gaining attention to reduce the ecological footprint associated with drone battery production and disposal.

In summary, batteries are the heart of drone technology, and their ongoing development is essential to unlocking new capabilities, improving reliability, and expanding the practical uses of drones across industries. The balance between energy capacity, weight, safety, and cost continues to drive research and innovation in this crucial field.

4.1.1 (LiPo) Batteries for Drones in Warfare

Lithium Polymer (LiPo) batteries have become the de facto power source for the majority of small unmanned aerial vehicles (UAVs), including hobby-grade drones repurposed for battlefield use. Their popularity stems from a unique balance of high energy density, lightweight design, flexible form factor, and rapid power discharge, which makes them particularly well-suited to the intense demands of aerial flight, especially in rotary-wing configurations.

In warfare contexts, LiPo-powered drones have found utility in a wide range of roles, including reconnaissance, target acquisition, direct engagement (e.g., explosive delivery), electronic warfare decoys, and logistical supply. Their ability to deliver short bursts of high power with minimal lag is critical in many of these mission profiles.

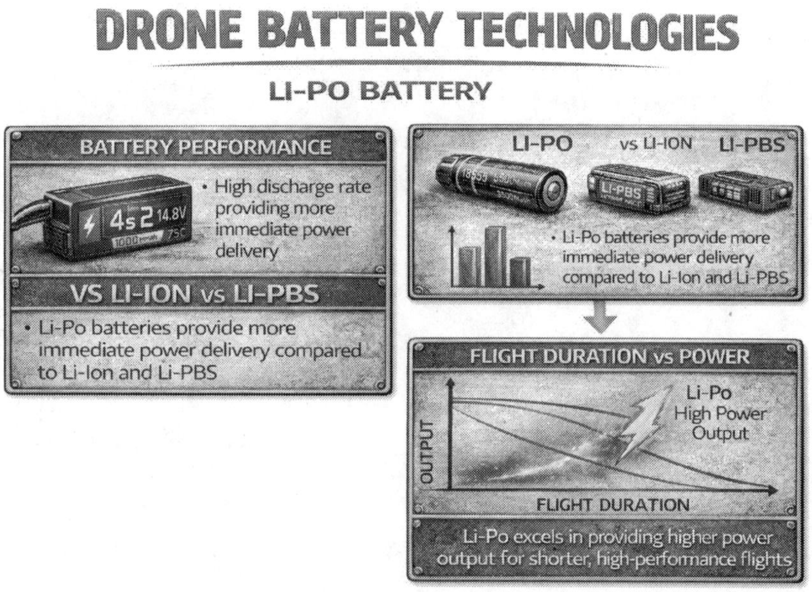

Yet, LiPo batteries are not without their drawbacks. Known for their volatility, short operational lifespan, and environmental sensitivity, they pose unique logistical and tactical challenges on the battlefield.

This section examines every aspect of LiPo batteries in the context of drone warfare—from core chemistry to tactical applications—providing a comprehensive understanding of their strengths, limitations, and evolving role in modern conflict.

4.1.2 (LiPo) Composition, Structure, Chemistry

Lithium Polymer (LiPo) batteries are a type of rechargeable lithium-ion battery that differs from traditional cylindrical lithium-ion cells primarily in their use of a solid or gel-like polymer electrolyte and a flexible pouch casing. The technical composition of a LiPo battery typically includes a cathode made of lithium cobalt oxide ($LiCoO_2$), a common choice for its high energy density, and an anode usually composed of graphite. Between these two electrodes lies a polymer-based electrolyte, often enhanced with liquid lithium salts to improve ionic conductivity. This semi-solid electrolyte is what gives the battery its "polymer" designation and allows for a more compact, flexible, and lightweight structure compared to rigid metal-cased lithium-ion batteries. The electrolyte acts as a medium for the transport of lithium ions during charging and discharging cycles, playing a crucial role in the battery's overall efficiency and safety.

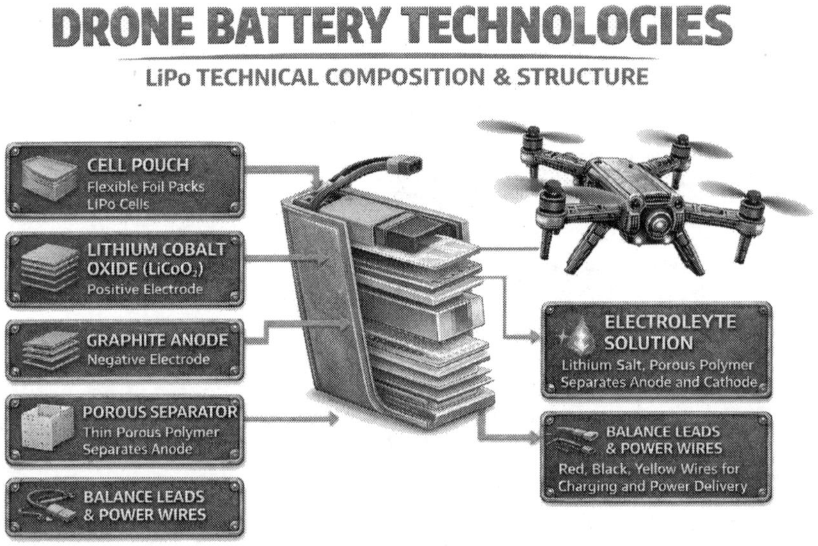

The strUcture of a LiPo battery is typically made up of multiple flat cells stacked together in a flexible, sealed foil pouch. This design provides several advantages, including reduced weight and greater shape versatility, which is especially valuable in drone applications where space is limited and weight savings are critical. Unlike cylindrical cells, LiPo cells can be manufactured in a wide range of shapes and sizes, allowing drone designers to optimise internal configurations for aerodynamics and balance. Each cell in a LiPo battery typically has a nominal voltage of 3.7 volts, and battery packs are commonly configured in series (e.g., 3S, 4S) to increase voltage or in parallel to increase capacity.

Chemically, LiPo batteries are similar in function to traditional lithium-ion batteries, operating on the movement of lithium ions from the anode to the cathode during discharge, and the reverse during charging.

However, the polymer-based electrolyte contributes to a thinner, lighter cell with lower risk of leakage. Despite these advantages, LiPo chemistry does come with certain trade-offs, including a higher sensitivity to overcharging, deep discharging, and physical damage. As such, they require precise battery management systems (BMS) to monitor cell voltage, balance charge between cells, and prevent overheating. Overall, the technical composition, structure, and chemistry of LiPo batteries make them ideal for drone warfare applications, providing a balance of energy density, weight efficiency, and form factor adaptability essential to high-performance, tactical UAVs.

4.1.3 (LiPo) Discharge Characteristics

One of LiPo's key strengths is its "high discharge rate", often labelled as the "C-rating." This rating determines how fast a battery can safely discharge its energy. For example, a 2200mAh 30C LiPo can theoretically deliver 66 amps continuously (2.2 × 30 = 66 A).

This capability is especially valuable for several types of drones. Quad-copters and hex-copters, for example, demand immense bursts of power during take-off and complex manoeuvres, making high-discharge batteries essential. FPV racing drones also rely heavily on rapid throttle response, where even a fraction of a second's delay can affect control and performance. In the case of combat drones, motor responsiveness and agility can make the difference between a successful mission and total failure—potentially even life or death for operators nearby. The Lithium Polymer (LiPo) battery's ability to rapidly discharge large amounts of energy makes it ideally suited for these high-performance rotary-wing drones, especially in the rugged and unpredictable conditions of combat environments.

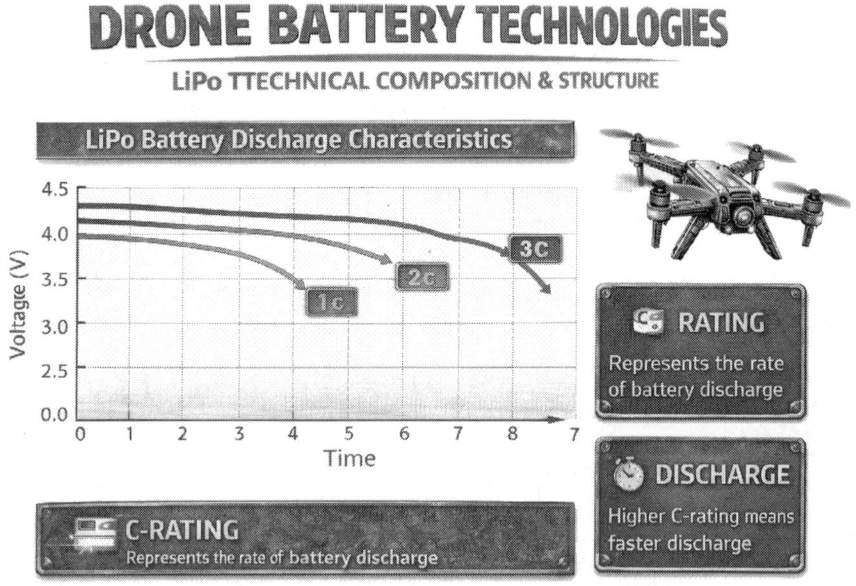

A crucial operational distinction lies in discharge rate. LiPo batteries often boast very high C-ratings, meaning they can deliver large amounts of current in short bursts—ideal for agile multi-rotors. Li-Ion batteries tend to have lower C-ratings, favouring endurance over burst power.

Comparison of Li-Po and Li-Ion Battery Technologies

Lithium Polymer (Li-Po) and Lithium-Ion (Li-Ion) batteries represent two dominant energy storage technologies used in modern drones, each optimised for different operational priorities. Li-Po batteries are characterised by their high discharge rates and lightweight, flexible pouch construction, making them ideal for applications requiring rapid power delivery. However, their energy density typically ranges between 150-250 Wh/kg, limiting overall endurance. Li-Po cells also exhibit a relatively short service life, commonly 150-300 charge cycles, and carry a high thermal and fire risk due to volatile electrolyte chemistry and susceptibility to mechanical damage or over-discharge.

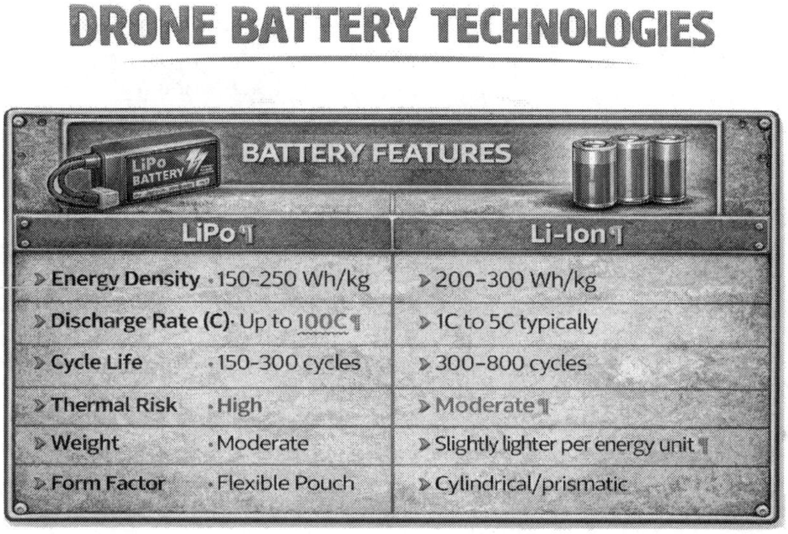

	LiPo¶	Li-Ion¶
Energy Density	150-250 Wh/kg	200-300 Wh/kg
Discharge Rate (C)	Up to 100C¶	1C to 5C typically
Cycle Life	150-300 cycles	300-800 cycles
Thermal Risk	High	Moderate¶
Weight	Moderate	Slightly lighter per energy unit¶
Form Factor	Flexible Pouch	Cylindrical/prismatic

In contrast, Li-Ion batteries prioritise energy capacity and longevity. With energy densities commonly between 200–300 Wh/kg, Li-Ion cells provide significantly greater stored energy for the same mass, enabling longer flight times. A key advantage is their substantially longer life cycle, often 300–800 cycles, more than double that of Li-Po batteries. Li-Ion chemistry is also more thermally stable, presenting a moderate thermal risk rather than the acute fire hazard associated with Li-Po packs. This improved safety profile comes at the cost of lower maximum discharge rates, making Li-Ion batteries less suited to high-current manoeuvring but ideal for endurance-focused missions.

Overall, Li-Po batteries favour performance and responsiveness, while Li-Ion batteries excel in endurance, safety, and long-term operational efficiency.

4.1.4 (LiPo) Advantages

Lithium Polymer (LiPo) batteries provide numerous advantages that make them highly suitable for drone warfare, particularly in tactical and short-range applications. One of their primary benefits is their high energy density relative to weight, typically around 150–250 Wh/kg. While this is lower than hydrocarbon fuels, it is significantly better than older battery types such as Nickel-Metal Hydride (NiMH) or lead-acid, making LiPo batteries far more efficient for powering drones. This higher energy density enables relatively long flight times while maintaining a lightweight payload, which is especially valuable in small- to medium-sized UAVs where weight is a critical design constraint.

Additionally, LiPo batteries support high discharge rates, allowing drones to accelerate rapidly and perform agile maneuvers—important in combat scenarios requiring quick evasion or precision strikes. Their compact and flexible form factor allows integration into a variety of drone shapes and sizes, improving overall aerodynamics and space efficiency. LiPo batteries can also be recharged quickly, supporting faster deployment and turnaround times between missions. Their lower cost and widespread availability make them ideal for mass-produced drones or swarm tactics, and their quiet electric operation offers a low acoustic signature, aiding stealth and reducing the risk of detection. Together, these features make LiPo batteries a practical and effective power source in modern drone warfare, ideal for fast, short, direct engagement.

4.1.5 (LiPo) Power Delivery Motor Response

LiPo batteries are highly valued in drone applications for their fast power delivery and ability to enhance motor responsiveness. Brush- motors, which are commonly used in drones, require rapid bursts of power to adjust their RPM during flight manoeuvres. LiPo batteries, thanks to their low internal resistance, can supply high current almost instantaneously, enabling drones to react quickly to control inputs. This rapid responsiveness is essential for maintaining hover stability in rotary-wing drones, executing evasive manoeuvres when under fire, and performing precision strikes that demand accurate positioning. In combat scenarios, where timing and agility are critical, LiPo technology provides a clear operational advantage through their direct motor response profile.

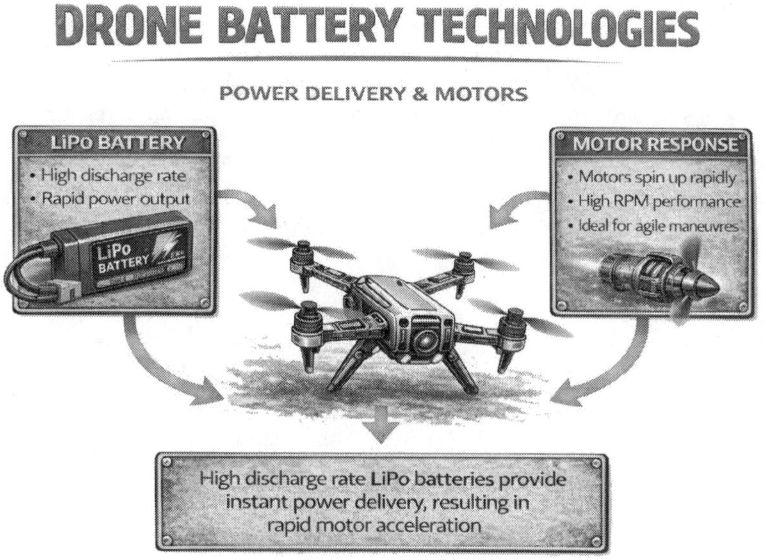

4.1.6 (LiPo) Ease of Access and Compatibility

LiPo batteries are readily available globally, often sold through RC hobby shops, electronics suppliers, or online marketplaces. For battlefield conditions, this makes them ideal for distributed operations where local sourcing may be necessary. Furthermore, most commercial drones use standardised connectors (e.g., XT60, Deans, JST), making swap-outs quick.

LiPo batteries also support swappable and modular design, a key feature in many modern drones. Quick-swap battery bays enable operators to carry multiple pre-charged LiPo packs and rapidly exchange them in the field, significantly reducing downtime. This capability supports high-tempo drone operations, particularly in fast-paced environments such as combat zones. It is especially advantageous for tactics involving drone swarms or rapid reconnaissance missions, where continuous deployment is critical. Additionally, LiPo-powered drones offer a low acoustic signature, making them far quieter than drones powered by internal combustion engines. This stealth advantage is vital in scenarios requiring discretion, such as reconnaissance missions behind enemy lines, covert drone strikes, and psychological operations (PSYOPS) where avoiding detection is essential to mission success.

4.1.7 (LiPo) Limitations and Challenges

Despite their advantages, LiPo batteries come with notable limitations that impact their effectiveness in extended combat operations. One of the primary drawbacks is their limited endurance and energy density. While LiPo batteries outperform older battery technologies in terms of power delivery, they fall significantly short when compared to fuels like gasoline, which offers around 12,000 Wh/kg, or hydrogen fuel at approximately 33,000 Wh/kg. This relatively low energy density restricts their use in long-range missions, especially for fixed-wing drones, where endurance is critical to mission success. On average, LiPo-powered rotary-wing drones achieve flight times between 10 and 30 minutes, while larger, more efficient fixed-wing drones may manage between 45 and 90 minutes. These short durations create tactical vulnerabilities, particularly in denied or contested airspace where drones may not have the luxury of returning to base for recharging or battery swaps.

In addition, LiPo batteries are chemically volatile and pose a serious safety risk in combat conditions. They are prone to thermal runaway, a chain reaction that can result in fire or explosion when batteries are punctured, overcharged, or exposed to excessive heat. In war zones, where drones may be damaged mid-flight or roughly handled during rapid deployment, the risk of fire increases significantly. As a result, many front-line units have adopted field-safe protocols, such as the use of fireproof battery bags, having sand or gel extinguishers available at launch points, and avoiding in-field charging where possible. These precautions, while necessary, add logistical complexity in already chaotic and resource-constrained environments. Another significant challenge is the temperature sensitivity of LiPo batteries. In cold climates —such as during the Ukrainian winter—LiPo batteries experience a sharp drop in performance, losing both capacity and voltage. Field tests and combat experience have shown that battery efficiency can fall by 30% or more under freezing conditions. Operators are often forced to warm batteries manually, using body heat or insulated pouches, and pre-plan shorter missions to account for diminished range. Spare batteries must be carried with thermal wraps to maintain viability. On the opposite end of the spectrum, extremely hot environments can lead to overheating, battery swelling, or chemical degradation. In these cases, teams must include cooling cycles, shaded storage, and heat-resistant containers as part of their operational planning.

Together, these limitations highlight the need for careful energy management, strategic mission planning, and continued innovation if LiPo-powered drones are to remain effective tools on the battlefield. While LiPo technology offers unmatched power-to-weight benefits in the short term, its endurance, safety, and environmental performance challenges must be addressed in any long-term drone warfare strategy.

4.1.8 (LiPo) Degradation and Life Cycle

LiPo batteries generally offer a lifespan of around 150 to 300 charge cycles under ideal conditions. However, in combat environments where batteries are subjected to intense use, rapid charging, and frequent cycling, their lifespan can drop significantly—sometimes falling below 100 effective cycles. As batteries degrade, several warning signs become apparent, including lower voltage under load, swelling due to gas buildup inside the cells, and noticeable reductions in flight time. These performance issues will compromise mission effectiveness and pose safety risks if not properly managed. As a result, effective inventory tracking and lifespan estimation become critical components of drone operations. Many units address this by labelling and rotating batteries, keeping detailed logs to ensure that only reliable packs are used during missions, thereby maintaining both safety and operational consistency in the field.

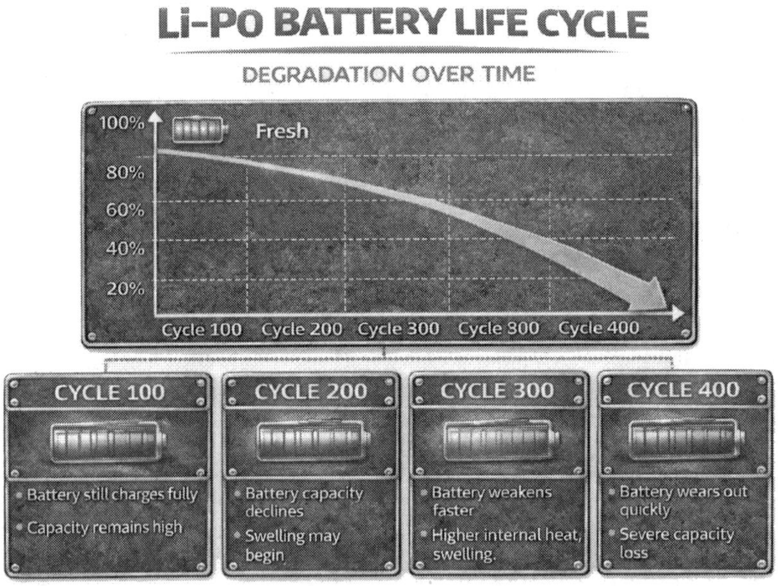

4.1.9 (LiPo) Recharging Infrastructure

Charging LiPo batteries in the field presents significant logistical challenges that impact operational effectiveness. Reliable recharging requires access to portable power sources, such as generators, vehicle inverters, or solar charging kits—each of which adds weight, complexity, and maintenance requirements to field deployments. Additionally, smart chargers with cell balancing capabilities are essential to prevent overcharging and ensure the safety and longevity of the batteries. Charging must also occur in fireproof or fire-resistant environments to mitigate the risk of thermal runaway, which can lead to fire or explosion if a battery fails. These requirements place a considerable burden on forward operating units, especially in fast-moving or resource-constrained environments. In prolonged engagements, the need to rotate batteries through limited charging infrastructure can create bottlenecks, ultimately restricting the drone sortie rate and reducing the overall tempo of aerial operations.

4.1.10 (LiPo) Summary, Lithium Polymer

In summary, LiPo batteries play a vital role in the effectiveness of combat drones, offering the high discharge rates, fast responsiveness, and compact form factor required for agile, high-performance aerial systems. Their ability to deliver rapid bursts of power makes them especially well-suited to rotary-wing platforms, FPV drones, and precision strike applications where motor agility and control are critical. Features such as swappable designs and low acoustic signatures enhance their value in dynamic combat environments, supporting quick redeployment and stealth operations. However, LiPo batteries come with important limitations—including limited endurance, sensitivity to temperature, chemical volatility, and complex charging requirements—that impose logistical and operational burdens on field units. Their relatively short lifespan, especially under combat stress, further necessitates careful inventory management and battery rotation. Despite these challenges, LiPo technology remains a cornerstone of modern drone warfare, providing a balance of performance, availability and flexibility that continues to shape how conflicts are fought on the front lines.

CHAPTER 5

Drone Battery Technologies (Li-Ion)

4.2.1 (Li-Ion) Batteries for Drones in Warfare

The Power Behind Aerial Warfare. The modern battlefield is undergoing a silent but significant transformation. While kinetic weapons, surveillance systems, and electronic warfare continue to evolve, one critical enabling factor often escapes attention: power storage technology. Among these, Lithium Polymer (LiPo) and Lithium-Ion (Li-Ion) batteries serve as the foundational energy source for a vast majority of small and medium-sized unmanned aerial vehicles (UAVs), including rotary-wing and fixed-wing hobby drones used in modern conflicts.

From the skies over Ukraine and Gaza to the mountain passes of Nagorno-Karabakh, drones have become a persistent and adaptable asset in reconnaissance, artillery spotting, tactical strikes, and even supply drops. Central to their flexibility is the lightweight, high-density power offered by advanced lithium-based batteries. While LiPo and Li-Ion batteries share similarities in chemistry and design lineage, they exhibit marked differences in application, performance, and handling, particularly under the strenuous demands of combat.

This book explores the scientific fundamentals, engineering trade-offs, battlefield performance, and strategic implications of both battery types. It offers guidance on how to choose, maintain, and deploy these power systems in rotary and fixed-wing drones operating in modern warfare environments.

4.2.2 (Li-Ion) Composition, Chemistry

Lithium-Ion (Li-Ion) batteries are a widely used rechargeable battery technology known for their high energy density, long cycle life, and reliable performance. Their technical composition includes three main components: a cathode, typically made from lithium metal oxides such as lithium cobalt oxide ($LiCoO_2$), lithium nickel manganese cobalt oxide (NMC), or lithium iron phosphate ($LiFePO_4$); an anode, usually composed of graphite; and a liquid electrolyte that facilitates the movement of lithium ions between the electrodes.

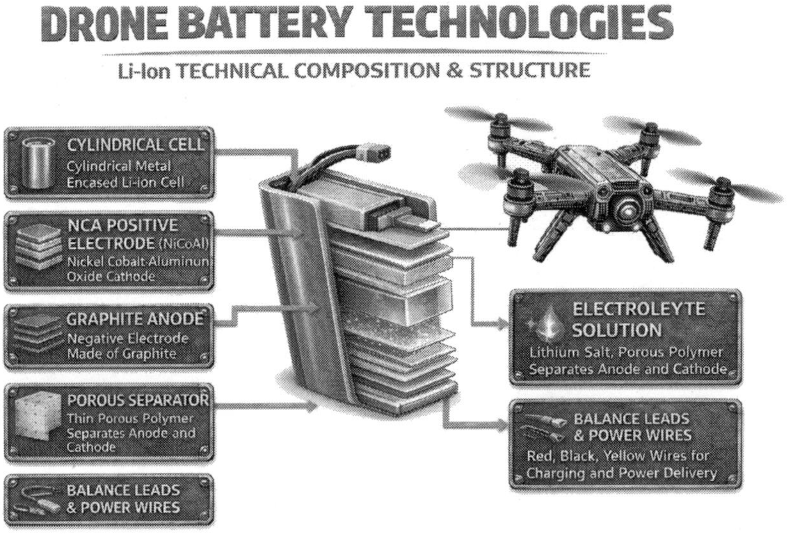

The electrolyte is typically a lithium salt such as $LiPF_6$ dissolved in a mixture of organic solvents like ethylene carbonate and dimethyl carbonate, chosen for their high conductivity and stability. A porous separator sits between the anode and cathode to prevent direct contact while allowing lithium ions to pass through. The working principle of Li-Ion batteries involves lithium ions migrating from the anode to the cathode during discharge, releasing energy, and then moving back to the anode during charging.

The structure of Li-Ion batteries generally follows a rigid cylindrical or prismatic format, housed in metal casings that provide mechanical strength and protection. This design is highly stable and suitable for applications where robustness and long-term durability are important, such as in electric vehicles, military equipment, and larger UAVs. Each Li-Ion cell typically has a nominal voltage of 3.6 or 3.7 volts and can be combined in series (to increase voltage) or in parallel (to increase capacity) to form battery packs tailored to specific power requirements. While Li-Ion batteries are often heavier and less flexible in shape compared to

Lithium Polymer (LiPo) batteries, they tend to offer superior energy density per volume and greater longevity, making them ideal for drones requiring long-endurance missions or energy-intensive payloads such as surveillance equipment and communication systems, including electronic warfare..

From a chemical standpoint, Li-Ion batteries offer high efficiency and relatively stable performance over hundreds or even thousands of charge cycles. They exhibit low self-discharge rates and are less prone to swelling or leakage than their LiPo counterparts. However, they are still susceptible to thermal runaway if overcharged, short-circuited, or damaged, necessitating the use of robust Battery Management Systems (BMS) to monitor cell voltage, temperature, and balance. In drone warfare, Li-Ion batteries are often chosen for missions requiring long flight times and consistent power delivery, thanks to their stable chemistry, strong structural integrity, and favourable energy density characteristics.

4.2.3 (Li-Ion) Discharge Characteristics

Lithium-Ion (Li-Ion) batteries exhibit discharge characteristics that are particularly favourable for drone applications where endurance, energy efficiency, and voltage stability are paramount. Understanding how Li-Ion batteries discharge under load is essential to optimising UAV performance, especially in military and surveillance roles where reliability and predictability are critical. One of the primary advantages of Li-Ion chemistry is its relatively flat discharge curve. Unlike older battery types such as NiMH or lead-acid, Li-Ion cells maintain a steady voltage for most of their discharge cycle, typically beginning at around 4.2 volts per cell when fully charged and gradually tapering down to approximately 3.0 volts when nearing depletion. The operational "sweet spot" lies between 3.7 and 3.3 volts, where the battery delivers most of its usable capacity. This flat voltage profile ensures that drone motors and onboard electronics receive consistent power throughout the majority of the flight, reducing performance drops and allowing for stable control, especially on long missions.

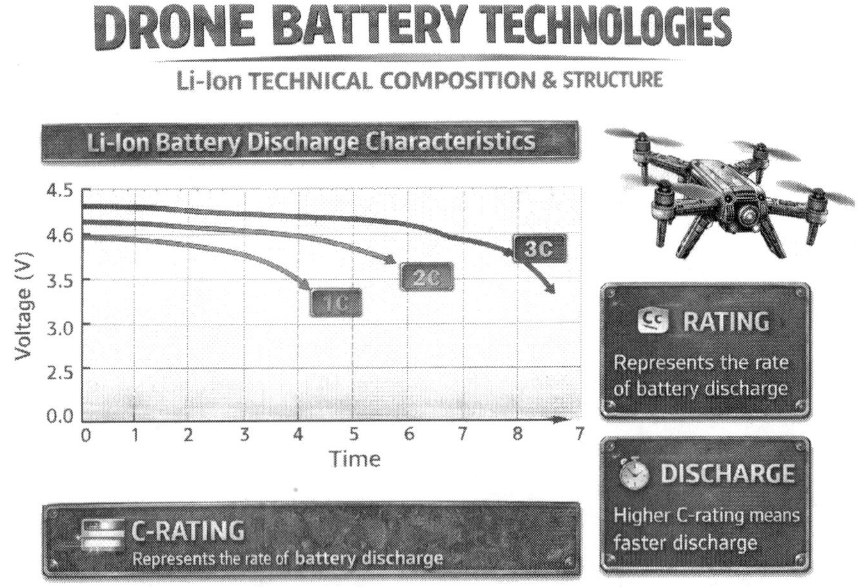

A critical element in evaluating discharge characteristics is the C rating, which refers to the rate at which a battery can safely be discharged relative to its total capacity. For instance, a 1C discharge rate means a battery can be discharged in one hour, delivering a current equal to its rated capacity (e.g., a 5000mAh battery at 1C provides 5 amps). Higher C ratings indicate the battery's ability to deliver more current in less time. While LiPo batteries often feature very high C ratings (20C to 60C or more) suitable for aggressive, high-thrust operations, typical Li-Ion cells have lower C ratings, usually between 1C and 5C. This means they are not designed for delivering large bursts of power but excel in scenarios where steady, moderate current draw is needed—such as in fixed-wing drones, long-range reconnaissance UAVs, or loitering munitions.

The implications of the C rating on drone design are significant. For drones that require longer flight durations with a low to moderate power draw—especially intelligence, surveillance, and reconnaissance (ISR) platforms—Li-Ion batteries offer superior energy density and better thermal management than LiPo batteries. Their ability to deliver a stable current over an extended period without significant voltage sag improves the overall efficiency of propulsion systems and onboard electronics. However, exceeding the recommended C rating can cause a rapid voltage drop, increased heat generation, and even permanent damage to the battery. For this reason, selecting the correct battery based on mission profile and power demand is essential. In high-load scenarios, designers may need to configure multiple Li-Ion cells in parallel to distribute current draw and prevent overheating or excessive voltage drop.

Depth of discharge (DoD) also plays a crucial role in Li-Ion battery performance and lifespan. These batteries can be safely discharged to around 80–90% of their total capacity without incurring long-term damage, but regularly draining below 3.0 volts per cell risks irreversible degradation and capacity loss. Consequently, modern drones are equipped with Battery Management Systems (BMS) that monitor voltage levels in real-time, ensuring that individual cells remain balanced and preventing over-discharge, which compromise mission integrity or cause total in-flight power failure. The BMS may also regulate discharge rates and shut off power if thresholds are exceeded, contributing to overall system safety.

Heat generation during discharge is another vital consideration. Li-Ion batteries generally produce less heat than LiPo cells under equivalent loads, but at higher discharge rates (closer to their maximum C rating), they can still heat up rapidly. Thermal buildup can affect both the battery and adjacent components in the drone's confined air-frame. Proper thermal management, such as passive cooling designs or airflow channels, is important when using Li-Ion batteries in high-temperature environments or during extended operations.

In conclusion, the discharge profile of Li-Ion batteries makes them particularly suitable for drone applications—both quad-copter and fixed-wing platforms—that emphasise efficiency, endurance, and voltage stability rather than extreme burst output. Their moderate C-ratings are well aligned with long-duration missions where consistent current draw and predictable power delivery outweigh the need for rapid throttle surges or high-thrust acceleration. Although they are less suited to high-performance racing drones or heavy-lift multi-rotors requiring sudden power spikes, Li-Ion cells have become a preferred option in military and strategic UAV systems, supporting persistent surveillance, extended-range strike capability, and long-distance communications with dependable electrical stability. A thorough understanding of C-rating limits, depth-of-discharge parameters, and thermal management behaviour remains essential to maximising their effectiveness in advanced UAV operations. As artificial intelligence and machine learning tools continue to refine mission analysis and performance modelling, deeper insights into battery optimisation and energy management will further enhance their application in drone warfare and next-generation unmanned systems.

4.2.4 (Li-Ion) Advantages

Lithium-Ion (Li-Ion) batteries provide a series of advantages that make them particularly well suited to drone warfare, especially in missions requiring extended endurance, dependable power delivery, and high energy efficiency. One of their most significant strengths lies in their high energy density, enabling substantial energy storage within a compact and lightweight structure. This makes them especially effective for fixed-wing platforms and long-endurance UAVs tasked with reconnaissance, surveillance, or communications relay, where sustained flight time is operationally decisive. Greater energy density directly translates into increased airtime, allowing drones to travel further distances or loiter over designated areas for prolonged observation, thereby expanding operational reach and enhancing strategic flexibility.

A further advantage is the relatively stable voltage output Li-Ion cells maintain throughout most of their discharge cycle. Unlike certain battery chemistry that experience steep voltage decline as capacity diminishes, Li-Ion batteries exhibit a flatter discharge curve, delivering predictable and consistent power. This stability is critical for maintaining flight control systems, onboard sensors, and communications equipment without performance fluctuation. In missions where timing, coordination, and precision are paramount—such as intelligence collection or target tracking—reliable voltage delivery reduces the risk of sudden system degradation. Their comparatively long cycle life also supports repeated charge and discharge cycles with limited capacity loss, contributing to cost-effectiveness and sustainability during prolonged operational campaigns.

Li-Ion batteries additionally offer strong volumetric energy efficiency, enabling more streamlined and aerodynamic air-frame designs without sacrificing capacity. Although their peak discharge rates are generally lower than Lithium Polymer (LiPo) alternatives, they remain well suited to platforms with moderate and steady power demands, including patrol UAVs, monitoring systems, and extended-range payload carriers. In such applications, where efficiency and duration outweigh high-thrust bursts, Li-Ion technology presents a balanced and reliable solution. Furthermore, their low self-discharge rate and broad environmental tolerance—when paired with modern Battery Management Systems (BMS)—enhance operational safety and longevity. BMS integration monitors cell health, regulates temperature, and prevents over-discharge, reinforcing reliability in austere environments. Collectively, these characteristics—endurance, voltage stability, operational lifespan, and efficiency—position Li-Ion batteries as a preferred energy source for a wide spectrum of unmanned systems, particularly within distributed swarm and mesh-networked environments on the evolving battlefield.

4.2.5 (Li-Ion) Power Delivery Motor Response

Lithium-Ion (Li-Ion) batteries are not only valued for their energy density and endurance but also for their ability to provide fast power delivery and responsive motor control—an increasingly important factor in drone warfare. Although traditionally seen as less capable of handling high current bursts compared to Lithium Polymer (LiPo) batteries, advances in Li-Ion cell technology have significantly improved their discharge capabilities. Many modern Li-Ion cells can now deliver moderate to high current outputs with sufficient speed to support sudden changes in drone motor demand. This enables reliable performance during dynamic maneuvers, sudden altitude adjustments, or rapid directional changes—crucial in combat environments where evasion, targeting, and stability under stress are vital.

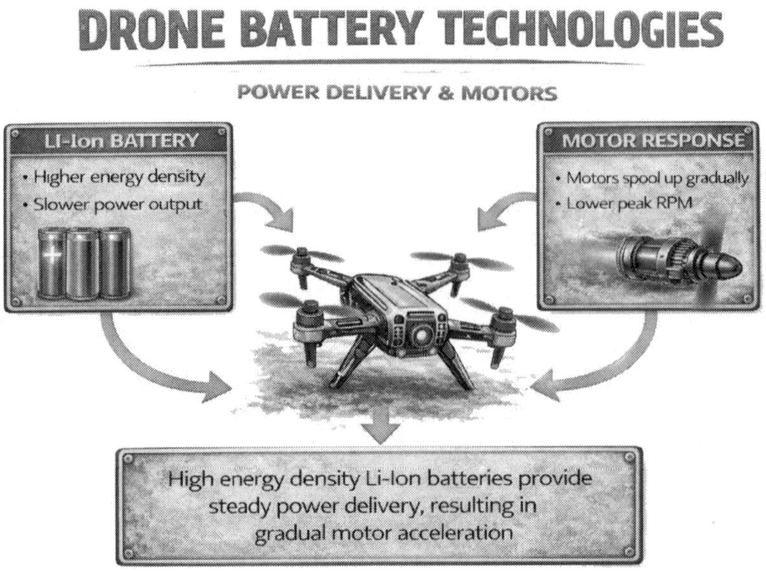

Fast power delivery in a drone context refers to how quickly the battery can respond to a power request from the flight controller and motors. Li-Ion batteries, thanks to their internal design and relatively low internal resistance, can deliver power with minimal delay, particularly when configured appropriately for parallel current distribution. While their peak discharge rates are lower than LiPo cells, they are more than capable of meeting the requirements of fixed-wing drones, hybrid UAVs, and even some multi-copters designed for endurance missions. For drones tasked with long-range surveillance or battlefield support, responsiveness doesn't necessarily mean explosive thrust—it means precise, stable motor control with dependable reaction to variable loads such as wind resistance, payload shifts, or terrain-following flight paths. Li-Ion batteries excel in providing consistent, on-demand power under such conditions.

Motor responsiveness is further enhanced by the stable voltage output characteristic of Li-Ion cells. The flatter discharge curve ensures that even as the battery depletes, the power supplied to motors remains relatively uniform. This prevents erratic behavior and ensures smooth flight dynamics across the entire mission duration. In contrast, batteries with steeper voltage drop-offs may cause sluggish motor response or early termination of mission-critical functions due to under-voltage cutoffs. In warfare, where drones may be required to change course quickly, lock onto moving targets, or reposition for real-time intelligence updates, the combination of fast power delivery and voltage stability can make the difference between mission success and failure.

Moreover, the reliability of Li-Ion batteries under controlled discharge rates contributes to overall system responsiveness by reducing thermal stress. Because they generate less heat under load compared to high-discharge alternatives, Li-Ion-powered systems can operate longer without requiring aggressive thermal management, reducing the risk of performance throttling or mid-air failure. The integration of smart Battery Management Systems (BMS) also allows real-time monitoring and optimisation of power output, ensuring that motor responsiveness is maintained even under fluctuating loads or partial battery drain. In sum, while Li-Ion batteries may not deliver the explosive thrust of LiPo batteries used in racing or strike drones, they provide a highly stable, responsive power source that meets the demands of precision UAV operations in modern combat scenarios. Their ability to deliver power quickly, maintain consistent voltage, and reduce thermal buildup ensures optimal motor responsiveness—an increasingly vital attribute in advanced drone design.

4.2.6 (Li-Ion) Ease of Access and Compatibility

Lithium-Ion (Li-Ion) batteries offer significant advantages in terms of ease of access and compatibility for drones, particularly in the context of modern warfare and commercial UAV operations. As one of the most widely used battery technologies in the world, Li-Ion batteries benefit from global manufacturing scale, standardisation, and a well-established supply chain. This makes them readily available across civilian and military markets, even in remote or resource-constrained environments. Drones that rely on Li-Ion batteries can often be powered by commonly available cells, including 18650, 21700, (NewBeeDrone Lion Bee) or prismatic formats, which are widely used in other technologies such as laptops, power tools, and electric vehicles. This cross-platform compatibility reduces logistical challenges and ensures a steady supply of replacements, spare parts, and charging infrastructure.

One of the key advantages of Li-Ion battery compatibility is their modular design. Many UAV systems, especially military-grade and commercial platforms, are now engineered with battery bays that accommodate standard Li-Ion cell packs or allow custom configurations using well-understood dimensions and electrical characteristics. This facilitates ease of integration for designers and maintainers alike, enabling quick field replacements, simplified repairs, and flexible upgrades. For example, fixed-wing surveillance drones and long-range mapping UAVs often use Li-Ion battery modules that can be hot-swapped or daisy-chained in parallel to extend range or mission duration. This versatility makes Li-Ion batteries a preferred choice for scalable drone platforms used in diverse mission profiles, from border patrol to disaster response and tactical reconnaissance including deep strike missions.

Charging compatibility is another major benefit. With Li-Ion technology, there is a broad ecosystem of chargers, balancing boards, and battery management systems (BMS) already in use across industries. This allows drone operators to re-purpose existing ground support equipment for multiple drone types, saving time and reducing the cost of deployment. The universal charging protocols and safety systems associated with Li-Ion batteries also improve operational readiness, as charging can occur with minimal risk under controlled conditions, and intelligent BMS can automatically regulate cell balancing, voltage limits, and thermal protection.

From a maintenance perspective, Li-Ion batteries also stand out for their lower degradation rate compared to some high-performance alternatives like LiPo. Their longer cycle life and resistance to swelling or puncture make them easier to handle, store, and transport—key factors in battlefield logistics. In environments where weight, space, and resilience are at a premium, Li-Ion batteries deliver reliable performance without the need for specialised storage cases or temperature controls. In summary, the ease of access, widespread compatibility, and adaptable infrastructure associated with Li-Ion batteries make them an ideal power source for drones in both military and civilian operations, providing logistical simplicity, operational flexibility, and strategic reliability across multiple platforms and mission profiles.

4.2.7 (Li-Ion) Limitations and Challenges

While Lithium-Ion (Li-Ion) batteries offer numerous advantages in drone warfare, they also come with several limitations and challenges that must be carefully considered when designing and deploying unmanned aerial systems. One of the most notable limitations is their relatively low peak power output compared to Lithium Polymer (LiPo) batteries. Li-Ion batteries typically have lower discharge rates (C-ratings), meaning they cannot deliver as much instantaneous current without risking voltage sag or overheating. This can be a disadvantage in high-thrust, high-manoeuvrability drones such as those used for rapid strikes, evasive maneuvers, or heavy payload delivery. In combat situations where responsiveness and burst power are essential, this limitation hinders drone performance or shortens mission capability.

Another significant challenge with Li-Ion batteries is their sensitivity to overcharging, deep discharging, and physical damage. While modern Battery Management Systems (BMS) mitigate some of these risks, the chemistry of Li-Ion cells remains inherently volatile under poor handling conditions. If a battery is overcharged or punctured—either during flight or due to a crash—it can enter thermal runaway, a dangerous condition that may result in fire or explosion. This is especially problematic in warfare environments, where drones may be exposed to enemy fire, shrapnel, or operational mishandling. As a result, additional protective engineering is often required to shield the battery compartment or to quickly isolate damaged cells from the rest of the system.

Thermal management is another challenge, particularly during long-duration flights or in hot climates. While Li-Ion batteries are more thermally stable than some alternatives, they still generate heat during discharge and can suffer from reduced performance or accelerated degradation when exposed to extreme temperatures. Without adequate cooling systems or ventilation, internal temperatures can rise quickly, leading to premature capacity loss or catastrophic failure. For drones operating in desert or tropical regions, this poses a real constraint on reliability and flight planning.

Storage and transport logistics also present operational hurdles. Despite being more stable than LiPo batteries, Li-Ion cells are still classified as hazardous materials and are subject to strict regulations for air and ground transport. This can complicate supply chains, especially for deployed forces needing rapid battery resupply. Furthermore, Li-Ion batteries lose capacity over time, even when not in use, and may require controlled storage conditions to prevent degradation during long periods of inactivity.

Finally, while Li-Ion batteries offer excellent energy density, they still fall short of hydrocarbon fuels in terms of total energy per kilogram. This inherently limits the range and payload of electric-powered drones compared to internal combustion or hybrid systems. In warfare, where endurance and load capacity are critical, this can restrict mission profiles unless compensated by larger battery arrays—adding weight and volume. In conclusion, while Li-Ion batteries are a practical and efficient power source for many drone warfare applications, they are not without technical and logistical drawbacks. Careful system design, thermal management, safety precautions, and mission planning are essential to mitigate these limitations and ensure operational effectiveness.

4.2.8 (Li-Ion) Degradation and Life Cycle

Lithium-ion (Li-Ion) drone batteries degrade over time through a combination of chemical, electrical, and mechanical processes that affect both their energy capacity and power delivery capability. From the first charge cycle, gradual aging begins as lithium ions move between the anode and cathode, forming and thickening the solid electrolyte interphase (SEI) layer on the anode. While the SEI is essential for stable operation, its continued growth consumes active lithium, leading to irreversible capacity loss over the battery's life. In drone applications, where high current draw is common, this process is accelerated compared to lower-demand consumer electronics.

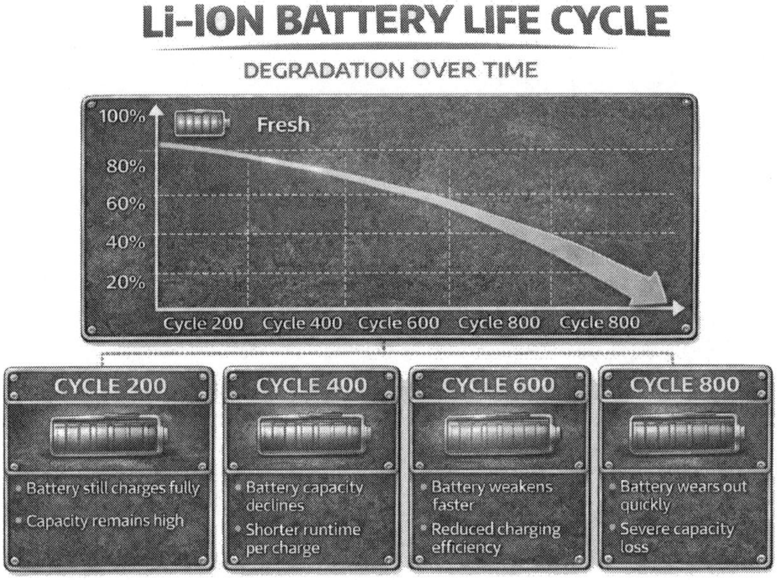

One of the primary drivers of Li-Ion degradation in drones is high discharge stress. Rapid throttle changes, sustained climb power, and aggressive manoeuvres increase internal temperatures and electrical resistance. Elevated heat accelerates electrolyte decomposition and promotes electrode material breakdown, particularly in compact UAV battery packs with limited thermal management. Environmental conditions further influence aging, with cold temperatures increasing internal resistance and hot climates accelerating chemical wear. Repeated exposure to voltage extremes—such as deep discharges below safe thresholds or frequent charging to maximum voltage—also shortens usable battery life.

Cycle life for Li-Ion drone batteries typically ranges from 300 to 800 full charge-discharge cycles, depending on cell quality, discharge rates, and operating conditions. As the battery ages, capacity retention declines gradually, but more critical for drones is the increase in internal resistance, which causes voltage sag under load. This results in reduced flight time, diminished peak power, and earlier low-voltage cutoffs during flight.

Even when a battery retains 70–80% of its nominal capacity, it may no longer be suitable for high-performance UAV operations.

End-of-life in Li-Ion drone batteries is defined not by complete failure, but by insufficient power delivery and increased safety risk. Swelling, excessive heat generation, and unstable voltage behaviour signal retirement. Proper charge management, moderate discharge rates, and temperature control can significantly extend Li-Ion battery life, making them a reliable but consumable component in modern drone systems.

4.2.9 (Li-Ion) Recharging Infrastructure

The recharging infrastructure required for Lithium-Ion (Li-Ion) batteries in warfare presents both strategic opportunities and operational challenges. One of the primary advantages of Li-Ion battery systems is their compatibility with widely available and standardised charging equipment. Unlike some speciality power sources, Li-Ion technology benefits from decades of civilian and military use, which has led to the development of robust, portable, and scalable charging solutions. In military drone operations, where speed, mobility, and logistics efficiency are paramount, having access to universal or modular charging stations enhances the flexibility and sustainability of missions. Field-deployable charging kits—often ruggedised and capable of charging multiple battery packs simultaneously—are already in use by military units, enabling rapid turnaround times between drone sorties.

Despite these advantages, establishing and maintaining Li-Ion recharging infrastructure in active conflict zones is not without its challenges. Power generation in the field is typically reliant on fuel-based generators, solar panels, or vehicle-mounted charging systems. While solar options offer a sustainable long-term solution with low operational signatures, they are heavily dependent on weather conditions and can be limited in energy output. Fuel-based generators, while more reliable, are noisy, require regular refuelling, and increase the logistical footprint of the operation—factors that can compromise stealth and mobility. In addition, extended drone operations involving larger Li-Ion battery packs necessitate high-capacity charging systems, which can further complicate field logistics and demand careful power management planning.

Once again, Battery Management Systems (BMS) also play a crucial role in the recharging process, as they regulate voltage, current, and temperature during charge cycles to ensure safety and efficiency. However, these systems require compatible charging hardware, and not all Li-Ion batteries use the same configuration or protocol. In warfare, this means that standardisation becomes critical. Units must ensure interoperability across various UAV platforms to prevent delays caused by incompatible connectors, chargers, or battery types. In large-scale operations involving mixed drone fleets, centralising and standardising recharging infrastructure can simplify operations and reduce the burden on logistics personnel.

Another issue is the time required to recharge Li-Ion batteries, which is significantly longer than re-fueling a combustion-powered drone. A full charge can take from 1 to 3 hours depending on the battery's capacity and the charger's power output. In high-tempo warfare environments, this downtime can become a bottleneck unless offset by using battery-swapping systems or maintaining large stockpiles of pre-charged batteries. The latter, however, requires significant space, weight, and cooling capacity—especially in hot climates where Li-Ion batteries degrade faster when stored at elevated temperatures.

In conclusion, while Li-Ion batteries offer a relatively mature and widely supported foundation for drone operations in warfare, their recharging infrastructure must be carefully planned and integrated into the overall logistics and mission design. Ensuring reliable power sources, fast and standardised charging systems, and safe storage protocols are essential to maintaining operational readiness and drone availability in the field.

4.2.10 (Li-Ion) Summary, Lithium-Ion

In conclusion, Lithium-Ion (Li-Ion) batteries have emerged as a cornerstone of modern drone warfare, offering a balance of energy efficiency, reliability, and technological maturity that aligns with the growing demands of unmanned aerial operations. Their high energy density allows for extended flight durations, making them ideal for missions involving surveillance, reconnaissance, target acquisition, communication relays and deep strike missions. Compared to older battery technologies like NiMH or lead-acid, Li-Ion batteries provide more power in a lighter, more compact form, which is vital for optimising drone payload capacity and flight endurance. While not as capable as Lithium Polymer (LiPo) batteries in terms of high burst current delivery, Li-Ion batteries excel in scenarios where steady, long-lasting energy is required, such as with fixed-wing drones or loitering munitions.

Their stable discharge curve and consistent voltage output ensure that drone systems remain responsive and reliable throughout the mission, enhancing control precision and reducing risks associated with sudden power loss or voltage drop-offs. Moreover, the widespread availability and standardisation of Li-Ion battery cells—such as the common 18650 or 21700 formats—offer logistical advantages in both civilian and military operations. This accessibility enables easier battery replacement, simplified recharging infrastructure, and greater compatibility across drone fleets and supporting equipment. In warfare scenarios where time, mobility, and supply chain resilience are critical, these factors greatly enhance mission effectiveness.

However, as highlighted, the use of Li-Ion batteries in drone warfare is not without challenges. Their limitations include lower maximum discharge rates compared to LiPo batteries, sensitivity to overcharging or deep discharging, thermal management concerns, and regulatory constraints on transportation and storage. These vulnerabilities necessitate careful engineering, including robust battery management systems (BMS), protective casing, thermal monitoring, and thoughtful logistics planning. Recharging infrastructure, though widely supported, requires dependable power sources in the field and introduces potential delays due to longer charge times compared to fuel-based systems. In high-intensity or rapid-deployment environments, this may necessitate the use of modular battery-swapping systems or pre-charged reserves to maintain operational tempo.

Ultimately, Li-Ion batteries represent a practical and effective power solution for a wide range of military drones. Their strengths in energy density, voltage stability, availability, and cycle life make them a go-to choice for endurance-based and precision-focused missions. As drone warfare continues to evolve, integrating smarter energy management systems, standardising battery configurations, and developing resilient recharging solutions will be key to fully leveraging the capabilities of Li-Ion technology.

CHAPTER 6

Drone Battery Technology (Li-PSB or Li-pSSB)

4.3.1 (Li-PSB) for Drones in Warfare

As drones continue to evolve into critical assets across both civilian and military sectors, the demands placed on their energy systems have become increasingly complex and demanding. Whether supporting long-endurance surveillance missions over hostile terrain or executing agile, high-intensity kinetic strikes in contested environments, drones are now expected to operate with precision, efficiency, and resilience under extreme conditions. At the heart of this capability lies the power source—arguably the most vital component of the drone after the motor and air-frame. Traditionally, Lithium Polymer (LiPo) batteries have dominated the UAV landscape, particularly in tactical rotary-wing platforms and hobbyist designs, due to their high power-to-weight ratio, compact form, and ability to deliver high discharge currents. Their quick responsiveness and relatively lightweight construction have made them an ideal choice for missions requiring short bursts of high thrust, manoeuvrability and front-line applications.

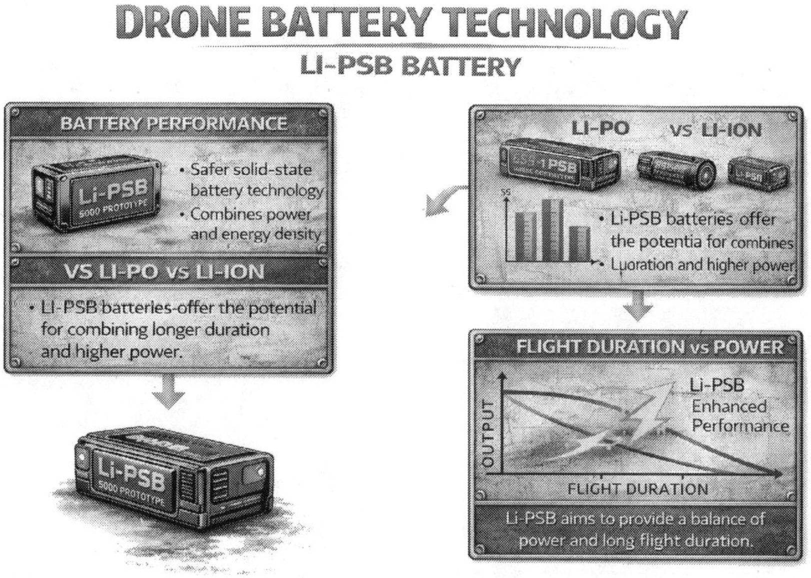

However, as operational requirements push the boundaries of conventional drone capabilities, the limitations of traditional LiPo chemistry are becoming more apparent. LiPo batteries, while powerful, carry inherent risks due to their flammable liquid electrolytes, thermal instability, and relatively short cycle life, especially under high-stress battlefield conditions. These vulnerabilities have prompted researchers, engineers, and defence planners to look beyond current battery technology and consider the next generation of power systems. At the forefront of this exploration is solid-state battery technology—specifically, solid-state lithium polymer batteries. This emerging class of batteries replaces the flammable liquid electrolytes used in traditional LiPo cells with solid-state or polymer-based electrolytes, resulting in a power source that is both safer and potentially more efficient.

Solid-state lithium polymer batteries promise several key advantages that align perfectly with the evolving needs of drone warfare. First and foremost, they offer significantly higher energy densities, allowing drones to fly longer and carry heavier payloads without increasing overall battery size or weight. This translates directly into extended mission duration, improved range, and enhanced payload versatility. Secondly, solid-state batteries exhibit far superior thermal stability compared to their liquid-electrolyte counterparts, reducing the risk of overheating, swelling, or fire—a crucial benefit in high-temperature combat zones or during prolonged flights. Additionally, their longer cycle life means fewer battery replacements are required, which reduces logistical strain and improves operational sustainability over time.

Perhaps most importantly for military applications, solid-state batteries enhance safety and survivability. In battlefield environments where damage to the aircraft is a real possibility, the reduced flammability and improved structural integrity of solid-state batteries could make the difference between mission failure and successful recovery. As drone warfare continues to mature, and as electronic systems become more sophisticated and power-hungry, the adoption of solid-state battery technology may represent the next critical frontier in drone power—enabling smarter, safer, and more capable UAVs on tomorrow's battlefield.

4.3.2 (Li-PSB) Composition, Structure, Chemistry

Solid-state batteries represent a significant advancement in energy storage technology, offering improved safety, energy density, and performance over conventional lithium-ion and lithium polymer batteries. Their technical composition, structure, and chemistry differ in several key areas from traditional batteries, most notably in the replacement of liquid electrolytes with solid materials. This fundamental change not only enhances the battery's safety profile but also unlocks new possibilities in design, efficiency, and thermal stability.

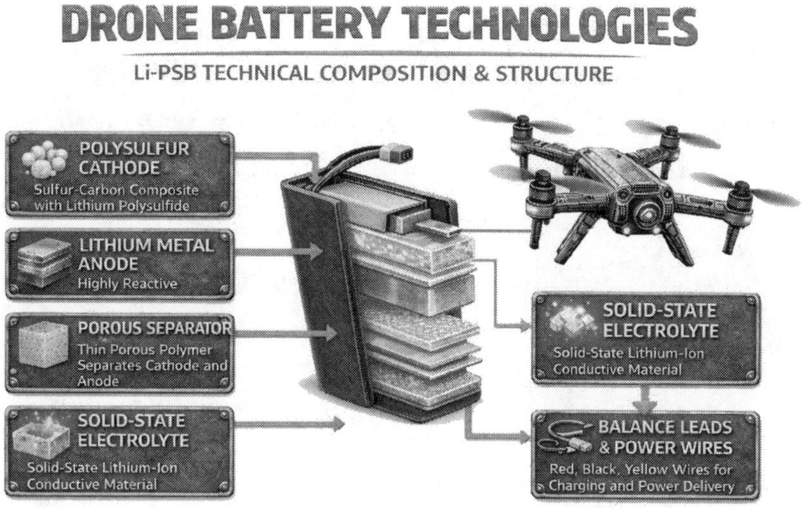

At the core of a solid-state battery is the electrolyte, which is composed of a solid material rather than the liquid or gel electrolyte found in conventional batteries. Solid electrolytes can be made from a variety of materials, including ceramics (such as lithium lanthanum zirconium oxide, or LLZO), sulfides, and solid polymers. These materials enable the conduction of lithium ions between the anode and cathode without relying on volatile or flammable solvents, significantly reducing the risk of leakage, fire, or thermal runaway. Solid electrolytes are also less prone to degradation and can function across a wider temperature range, which is particularly advantageous in high-performance applications such as drone warfare, where batteries may be exposed to extreme environmental conditions.

The structure of solid-state batteries also allows for more compact and efficient designs. Typically, the battery consists of a dense, multi-layer structure where the solid electrolyte is sandwiched between a lithium metal anode and a high-voltage cathode, such as lithium nickel manganese cobalt oxide (NMC) or lithium iron phosphate (LFP). The use of lithium metal as the anode, which is made possible by the solid electrolyte, dramatically increases energy density. This is because lithium metal has a much higher specific capacity compared to the graphite anodes used in conventional lithium-ion batteries. The compact stacking of solid components allows for thinner, more space-efficient battery designs, which can directly translate into lighter drones with longer flight times or higher payload capacities.

Chemically, solid-state batteries operate on the same fundamental principle as other lithium-based batteries: the movement of lithium ions from the anode to the cathode during discharge and back during charging. However, the solid-state electrolyte alters the dynamics of ion movement and often requires specific design considerations to maintain good ionic conductivity at room temperature. Innovations in material science, such as the development of doped ceramics and hybrid electrolytes, have improved conductivity and reduced interfacial resistance between layers.

In conclusion, the technical composition, structure, and chemistry of solid-state batteries offer significant advantages in terms of energy density, safety, and durability. These characteristics position solid-state batteries as a promising power solution for next-generation drones, where performance, compactness, and operational reliability are increasingly critical.

4.3.3 (Li-PSB) Discharge Characteristics

Lithium Polymer Solid-State Batteries (Li-PSBs) are an emerging power source that offer distinct discharge characteristics well-suited for the evolving needs of drone warfare and high-performance unmanned aerial vehicles (UAVs). Unlike traditional Lithium Polymer (LiPo) or Lithium-Ion (Li-Ion) batteries, Li-PSBs replace the flammable liquid electrolyte with a solid or semi-solid polymer-based material. This change not only enhances safety but also influences the battery's discharge behaviour, current-handling capacity, and overall performance under load.

One of the defining discharge characteristics of Li-PSBs is their ability to maintain voltage stability under higher current loads. Traditional batteries often suffer from voltage sag—where the voltage drops quickly during high-power demand—especially near the end of the discharge cycle. Li-PSBs, however, tend to maintain a flatter discharge curve, offering more consistent voltage delivery across a wider state of charge. This characteristic is particularly important in drone operations, where stable power is critical for maintaining flight control, motor efficiency, and reliable operation of onboard electronics.

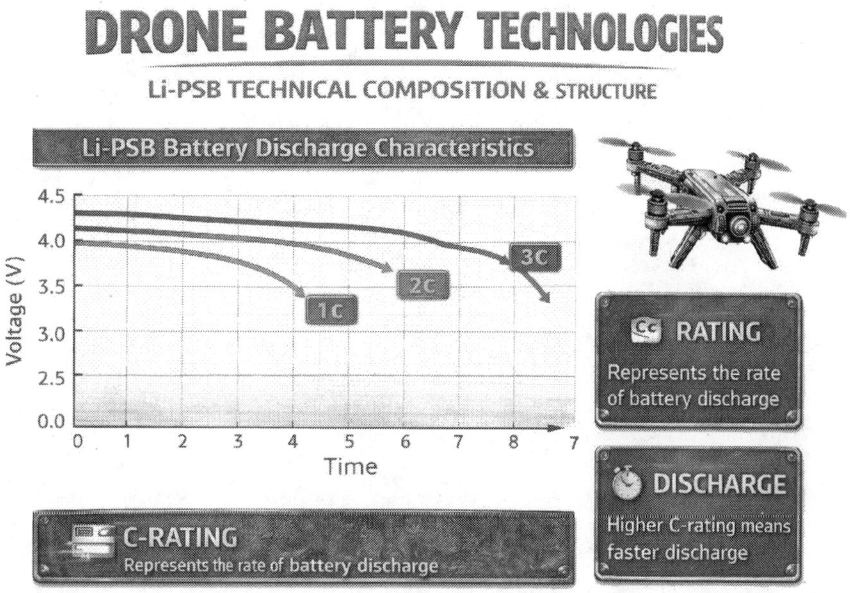

The C rating of a Li-PSBs defines how quickly the battery can safely discharge its energy. For example, a 5,000mAh (5Ah) battery with a 10C discharge rating can deliver 50 amps of continuous current. Li-PSBs are expected to offer C ratings comparable to or even exceeding those of conventional LiPo batteries, with the added benefit of thermal stability and reduced risk of combustion under stress. Some experimental and early commercial Li-PSBs report C ratings in the range of 10C to 20C for continuous discharge, and even higher for short bursts, making them viable for drones requiring both endurance and power surges—for instance, during vertical takeoff, evasion maneuvers, or payload delivery.

Li-PSBs also handle high discharge rates more efficiently due to their solid-state structure, which minimizes internal resistance and heat buildup. Unlike liquid-based electrolytes, solid-state materials are less likely to degrade under repeated high-load cycles, improving cycle life even when the battery is regularly subjected to aggressive discharge patterns. This makes Li-PSBs suitable for drones operating in combat zones, where quick redeployment and extended sortie counts are necessary.

Another advantage is thermal behaviour during discharge. While high C-rate discharges can generate significant heat in conventional batteries, Li-PSBs exhibit better thermal management characteristics. The solid electrolyte does not expand or degrade as rapidly under heat, reducing the risk of swelling, venting, or fire. This allows for safer operation under demanding flight conditions, especially in high-temperature environments or when operating enclosed drone designs with limited airflow.

DRONE BATTERY COMPARISON

LiPo¶	Li-Ion¶	Li-Ion¶	Li-PSB¶
Energy Density	150–250 Wh/kg	200–300 Wh/kg	200–400 Wh/kg
Discharge Rate (C)	Up to 100C ¶	1C to 5C typically	<1C at present
Cycle Life	150–300 cycles	300–800 cycles	>500 cycles (future potential)
Thermal Risk	High ¶	Moderate ¶	Low ¶
Weight	Moderate	Slightly lighter per energy unit ¶	
Form Factor	Flexible Pouch	Cylindrical/prismatic	
Form Factor	Flexible Pouch	Rigid or semi-flexible	

In summary, the discharge characteristics of Lithium Polymer Solid-State Batteries—marked by high and stable C ratings, minimal voltage sag, and excellent thermal performance—position them as a strong candidate for next-generation drone platforms. Their ability to deliver high current safely and consistently aligns with the increasing demands of modern UAV missions, where both performance and safety are non-negotiable.

4.3.4 (Li-PSB) Advantages

Lithium Polymer Solid-State Batteries (Li-PSBs) represent a meaningful evolution in energy storage technology, offering distinct advantages for advanced UAV applications. By combining the structural flexibility of traditional Lithium Polymer (LiPo) cells with the enhanced safety characteristics of solid-state systems, Li-PSBs replace the flammable liquid electrolyte found in conventional LiPo or Li-Ion batteries with a solid or semi-solid polymer-based alternative. This transition significantly improves thermal stability, reduces fire risk, and enhances overall energy efficiency. In operational environments where drones are exposed to physical stress, temperature extremes, and sustained high-demand missions, these characteristics provide measurable performance and survivability benefits.

A major strength of Li-PSBs lies in their improved safety profile. Conventional lithium batteries remain vulnerable to thermal runaway under high discharge loads, puncture, or impact damage—conditions not uncommon in demanding operational settings. In contrast, the non-flammable solid electrolyte within Li-PSBs greatly reduces the likelihood of ignition, while their more rigid internal structure resists swelling and mechanical compromise. This resilience allows UAV platforms to operate with increased reliability in harsh environments where heat, vibration, or structural shock could otherwise degrade battery integrity.

Li-PSBs also offer higher energy density compared with many traditional LiPo configurations, enabling greater energy storage within equal or reduced volume. This improvement supports longer flight durations, expanded mission radii, or the integration of additional onboard systems without proportionally increasing air-frame mass. For long-endurance reconnaissance drones, loitering platforms, or persistent surveillance systems, such gains in energy efficiency translate directly into enhanced operational reach. Extended flight windows further support advanced AI-driven analytics and machine learning model development, as longer missions enable broader data acquisition and more comprehensive performance modelling.

Beyond storage capacity, Li-PSBs deliver strong power output characteristics, including high C-rate discharge capability. This allows rapid and consistent current delivery for responsive motor performance, improved acceleration, and precise manoeuvring—attributes essential for evasive action or rapid repositioning. Notably, these batteries generate less heat and maintain more stable voltage under load compared to liquid-electrolyte counterparts, contributing to improved system durability. Their extended cycle life and reduced degradation also enhance long-term sustainability, supporting repeated deployment cycles with minimal performance loss. Collectively, these attributes position Lithium Polymer Solid-State Batteries as a promising power solution for next-generation UAV systems requiring endurance, safety, and operational resilience.

4.3.5 (Li-PSB) Power Delivery Motor Response

Lithium Polymer Solid-State Batteries (Li-PSBs) offer significant improvements in fast power delivery and motor responsiveness, making them highly attractive for drone applications, especially in tactical and high-performance warfare scenarios. These batteries combine the lightweight, high-output characteristics of traditional lithium polymer cells with the enhanced safety, thermal stability, and efficiency of solid-state electrolyte technology. The result is a power source capable of delivering quick bursts of energy without sacrificing stability, making Li-PSBs ideal for UAV missions that demand rapid responsiveness and high agility, such as evasion maneuvers, strike missions, and payload drops.

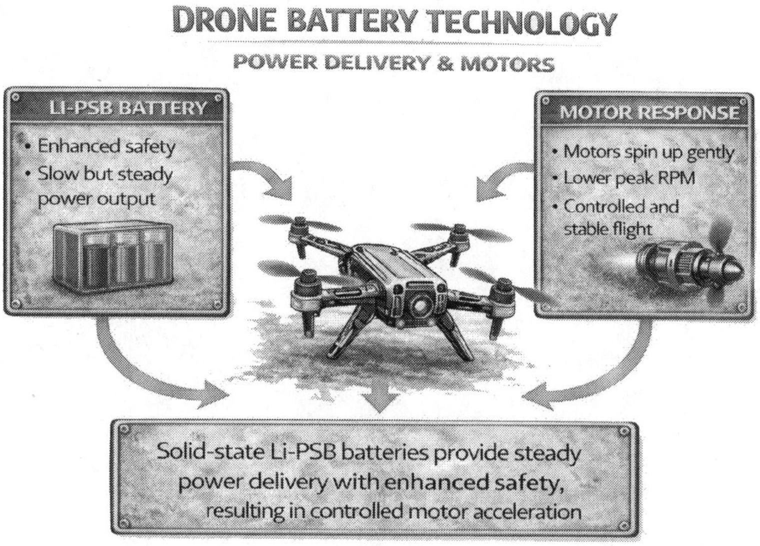

One of the standout features of Li-PSBs is their ability to deliver high discharge currents without compromise, often reflected in elevated C-ratings that support fast and sustained energy output. In drone motors, this translates directly into improved throttle response and better handling of dynamic flight demands. Whether the UAV is accelerating rapidly, making sharp directional changes, or climbing vertically with a heavy payload, the battery's ability to deliver consistent power without delay ensures that the motors respond immediately and efficiently. This level of responsiveness is critical in combat zones, where drones may need to evade enemy fire, respond to shifting mission objectives, or navigate unpredictable terrain in real-time.

What sets Li-PSBs apart from conventional LiPo or Li-Ion batteries is the reduced internal resistance offered by their solid-state design. Lower resistance means less energy is lost as heat during high discharge cycles, allowing more power to reach the motors quickly. This not only boosts performance but also reduces the thermal burden on the system, extending the life of both the battery and the drone's internal electronics. Additionally, the solid-state electrolyte's structural integrity prevents issues such as swelling or electrolyte leakage under stress—problems that can impair performance or cause catastrophic failure in traditional batteries during high-power draw situations.

The consistent voltage output of Li-PSBs across a broad range of discharge states also contributes to stable motor operation. In contrast to batteries that suffer from voltage sag under load, Li-PSBs maintain a flatter discharge curve, ensuring that the motors continue to receive a reliable power supply even as the battery nears depletion. This means drones can maintain full manoeuvrability and propulsion performance throughout the flight, rather than becoming sluggish or unstable toward the end of a mission. For autonomous or remotely piloted UAVs operating in complex or contested airspace, this reliability is essential for mission success, resistance to electronic warfare and range capability.

In summary, the fast power delivery and motor responsiveness enabled by Li-PSBs represent a significant step forward in UAV performance. With high discharge capability, voltage stability, and superior thermal handling, these batteries provide the responsive power systems needed for advanced drone warfare, enabling faster motor reaction times, sharper control, and more effective deployment in dynamic and dangerous environments.

4.3.6 (Li-PSB) Ease of Access and Compatibility

Polymer Solid-State Batteries (PSBs) are gaining attention in the drone industry due to their superior safety and performance characteristics, but ease of access and compatibility are still evolving factors in their adoption. Unlike traditional Lithium Polymer (LiPo) and Lithium-Ion (Li-Ion) batteries, which benefit from decades of widespread commercial use and standardisation, PSBs are still emerging technologies. As such, their accessibility in the current market is limited, with relatively few manufacturers producing drone-ready PSB packs in standardised sizes. This can make sourcing replacement batteries or integrating them into existing UAV platforms more challenging at present. However, as demand grows and production scales up, PSB availability is expected to increase significantly, leading to broader adoption and improved ease of access across both civilian and military drone sectors.

Despite these limitations, PSBs show considerable promise for future compatibility in UAV systems. Their physical format can often be tailored to mimic existing LiPo or Li-Ion battery designs, allowing for retrofitting into current drone configurations with minimal modification. This flexibility in form factor means PSBs can be developed to match common voltage and capacity requirements, ensuring that they remain interoperable with a wide variety of drones, from lightweight quad-copters to long-endurance fixed-wing platforms. Additionally, as battery management systems (BMS) evolve to accommodate PSB-specific chemistry and discharge profiles, seamless integration into existing drone architectures is becoming increasingly feasible.

Another factor contributing to the future compatibility of PSBs is their stable voltage behavior and consistent discharge characteristics. These features make them predictable and easier to integrate with existing drone electronics, which are often calibrated to expect certain voltage ranges and response times. As BMS software adapts to handle PSB-specific parameters—such as different charging curves or thermal thresholds—compatibility with existing drone firmware and flight controllers will improve, further lowering the barrier to adoption.

Moreover, PSBs offer logistical benefits in terms of transport and storage. Their solid-state composition means they are far less likely to catch fire, swell, or leak, making them easier to handle and ship under strict regulatory conditions—an important consideration for military logistics and international drone operations. In conclusion, while PSBs are not yet as accessible as older battery technologies, their increasing compatibility, adaptable design, and safer handling profile position them as a highly promising solution for the next generation of drone platforms. As production and standardisation progress, PSBs are expected to become a common power source, delivering the same ease of access and versatility that LiPo and Li-Ion batteries currently offer.

4.3.7 (Li-PSB) Limitations and Challenges

Despite their many advantages, Lithium Polymer Solid-State Batteries (Li-PSBs) also present a number of limitations and challenges that currently restrict their widespread adoption in drone warfare and commercial UAV applications. While their enhanced safety, energy density, and thermal stability offer compelling benefits, the technology remains in a relatively early stage of development, facing hurdles in manufacturing, cost, scalability, and integration.

One of the primary challenges is manufacturing complexity and cost. Producing Li-PSBs involves more intricate fabrication processes than traditional Lithium-Ion or Lithium Polymer batteries. The materials used in solid electrolytes, such as ceramic or polymer composites, often require precise conditions for purity, pressure, and layering. This adds significant cost to the production line and limits the number of manufacturers capable of delivering high-performance, drone-compatible PSBs at scale. As a result, Li-PSBs remain more expensive per unit than their liquid-based counterparts, making them less accessible to small UAV manufacturers or militaries operating with budget constraints.

Material compatibility and interface stability also pose significant technical obstacles. In solid-state designs, maintaining strong contact between the solid electrolyte and the electrode materials is critical for efficient ion transfer. However, mismatches in thermal expansion, interface degradation, or mechanical stress during charge and discharge cycles can lead to increased resistance and reduced battery life. These interface issues must be carefully engineered, often requiring novel material combinations or advanced coatings, which can further complicate the design and increase development time.

Another limitation is reduced performance at low temperatures. Many current solid-state electrolytes exhibit lower ionic conductivity in cold conditions compared to liquid electrolytes. This will impair drone functionality in high-altitude or cold-weather operations, reducing both discharge capacity and responsiveness. While research is ongoing to improve the low-temperature performance of PSBs, such as through doped polymers or hybrid electrolyte systems, this remains a concern for military drones operating in diverse environments.

Charging times and energy transfer rates also present a bottleneck. While Li-PSBs can support relatively high discharge rates, their charging speeds are often slower than expected due to limitations in ion mobility across solid-state materials. Without advanced thermal management and precise current control, fast charging can lead to internal stress, reduced cycle life, or performance degradation. This limits rapid redeployment of drone units in high-tempo warfare scenarios unless additional pre-charged batteries are kept in reserve.

Finally, standardisation and integration are still evolving. Because PSBs are not yet widely adopted, many UAV systems are not natively designed to accommodate them, requiring custom configurations or modified battery management systems (BMS) to ensure compatibility and safety. Until industry standards are more firmly established, widespread deployment of Li-PSBs will require time and investment. In summary, while Li-PSBs offer a promising leap forward in drone battery technology, addressing their technical, logistical, and economic challenges will be critical to realising their full potential on the battlefield and beyond.

4.3.8 (Li-PSB) Degradation and Life Cycle

Lithium polymer solid-state batteries (Li-PSB) represent an emerging energy storage technology with degradation and life-cycle characteristics that differ meaningfully from conventional liquid-electrolyte Li-Ion and Li-Po cells. In Li-PSB systems, the liquid electrolyte is replaced by a solid or semi-solid polymer electrolyte, which significantly alters aging mechanisms. One of the key advantages is the suppression of electrolyte evaporation, leakage, and gas formation, all of which are major contributors to degradation in traditional drone batteries. As a result, Li-PSB cells generally exhibit lower self-discharge rates, improved structural stability, and slower calendar aging when stored under controlled conditions.

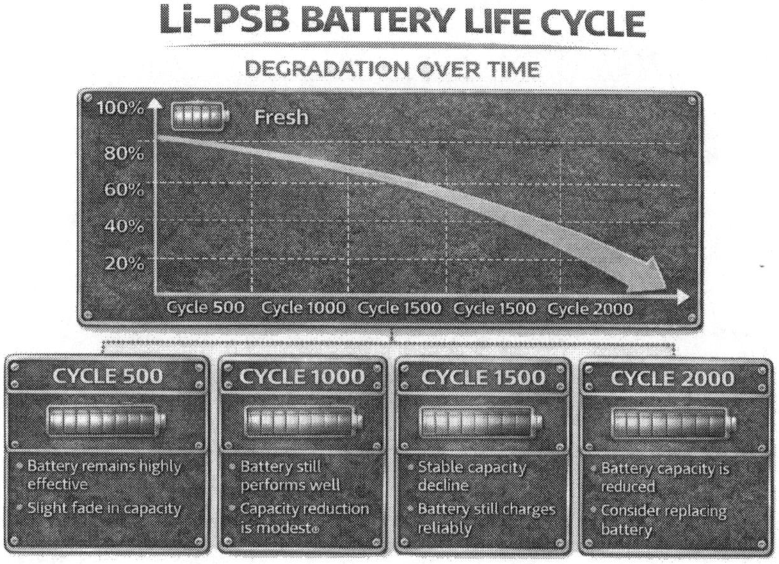

Degradation in Li-PSB batteries is primarily driven by interfacial resistance growth between the solid electrolyte and electrode materials. Over repeated charge-discharge cycles, microscopic mechanical stresses develop due to electrode expansion and contraction, which can degrade contact quality at these interfaces. This leads to gradual increases in internal resistance and reduced power delivery, though typically at a slower rate than liquid-electrolyte batteries. Unlike Li-Ion cells, Li-PSB designs greatly reduce the risk of lithium dendrite formation, extending usable cycle life and improving safety during high-load drone operations.

Thermal stress remains an important factor in Li-PSB aging, but solid electrolytes are generally more tolerant of elevated temperatures and less prone to rapid decomposition. Drone operations involving sustained high current draw still accelerate wear, particularly in propulsion-heavy multi-rotor platforms, yet degradation tends to be more linear and predictable. Current Li-PSB prototypes demonstrate projected cycle lives of 800 to over 1,500 cycles under moderate discharge rates, substantially exceeding conventional UAV battery technologies. A significant improvement.

End-of-life characteristics in Li-PSB batteries are marked by reduced peak power rather than abrupt capacity loss. For drone applications, this means batteries may remain energy-dense but become unsuitable for high-thrust manoeuvres as resistance rises. While commercial availability remains limited, Li-PSB technology offers significant potential for longer service life, enhanced safety, and improved reliability in future drone energy systems, particularly for long-endurance and mission-critical UAV platforms.

4.3.9 (Li-PSB) Recharging Infrastructure

The recharging infrastructure for Lithium Polymer Solid-State Batteries (Li-PSBs) in drone applications presents a developing area of technological support that combines both opportunity and complexity. As this next-generation battery technology gains traction in military and commercial UAV systems, its unique requirements necessitate adjustments in how drones are powered, recharged, and maintained in the field. Unlike traditional Lithium-Ion or LiPo batteries, which benefit from mature, widely available charging systems, Li-PSBs require more specialised recharging protocols and hardware to match their unique electro-chemical behaviour and solid-state architecture.

A key distinction of Li-PSBs lies in their charging profile. Solid-state batteries typically exhibit different voltage and current tolerances compared to conventional batteries. Their charge acceptance rate is influenced by the nature of the solid electrolyte, which, while offering improved safety and thermal stability, may impose limitations on how quickly ions move during charging. This results in longer charging times in many designs, especially without optimised thermal and electrical management systems. Fast charging may be possible in future iterations of the technology, but doing so safely requires intelligent chargers capable of monitoring and adjusting to the battery's condition in real time. This adds to the complexity of the ground support systems needed to maintain drone fleets using PSB technology.

In field operations, particularly in military settings, this translates into a need for advanced mobile charging stations. These must be rugged, lightweight, and capable of safely managing PSB-specific charging cycles. Furthermore, integration with existing UAV ground infrastructure is not yet universal, meaning that units transitioning to PSBs may require additional investment in compatible charging tools and connectors. Standard LiPo or Li-Ion chargers are often not suitable for PSBs due to differences in voltage thresholds and the risk of overcharging or undercharging, which could reduce battery life or even compromise safety.

That said, PSBs also offer advantages when it comes to charging logistics. Their superior thermal stability and reduced fire risk mean that they can be charged more safely in confined spaces or near sensitive equipment. This is particularly valuable in battlefield or mobile command post environments, where space is limited and safety is paramount. Additionally, PSBs can be charged at more varied environmental temperatures, allowing for more flexible deployment in diverse terrains without the need for intensive thermal management during recharging.

In summary, the recharging infrastructure for Li-PSBs is still maturing and must be tailored to the battery's specific electro-chemical characteristics. While the technology introduces some new challenges, particularly in terms of specialised equipment and compatibility, it also paves the way for safer, more resilient charging operations in both military and civilian drone applications. As adoption increases and industry standards emerge, the supporting infrastructure for PSBs is expected to become more accessible, standardised, and efficient.

4.3.10 (Li-PSB) Summary, Li-Solid-State

In summary, drone batteries play a pivotal role in shaping the capabilities, limitations, and future direction of unmanned aerial systems across both civilian and military domains. As drones become increasingly essential for tasks ranging from surveillance, mapping, and logistics to high-stakes combat missions, the power source that drives them becomes one of the most critical components. Battery performance directly influences a drone's flight time, payload capacity, speed, manoeuvrability, and operational reliability. The choice of battery technology—whether traditional Lithium-Ion (Li-Ion), Lithium Polymer (LiPo), or emerging innovations like Lithium Polymer Solid-State Batteries (Li-PSBs)—is central to the strategic effectiveness and efficiency of drone missions and future development.

Li-Ion batteries have proven themselves reliable for long-duration, steady-performance tasks, offering high energy density and long cycle life. Their widespread availability and mature infrastructure make them a practical choice for many drone operations, especially where endurance and efficiency are key. However, they come with limitations in thermal stability and discharge rates, which restrict their use in high-thrust or rapid-response tactical drones. LiPo batteries, on the other hand, have dominated the realm of performance-oriented UAVs due to their high discharge capabilities and excellent power-to-weight ratio. Their ability to deliver quick bursts of power makes them suitable for drones that require agility, fast climb rates, or heavy-lift capability. Yet, their volatility, shorter lifespan, and sensitivity to damage pose risks, particularly in military environments where durability and safety are paramount.

Emerging technologies like Li-PSBs offer a promising fusion of high performance and safety. These batteries address many of the shortcomings of conventional chemistry, particularly in terms of thermal management, energy density, and mechanical resilience. Their solid-state design reduces the risk of fire or explosion, enhances performance stability, and potentially extends cycle life. However, their integration into the drone sector is still in early stages, with challenges related to cost, standardisation, charging infrastructure, and manufacturing scalability that need to be overcome before widespread adoption is feasible.

Ultimately, the evolution of drone battery technology will continue to be a key enabler of progress in UAV design and mission capability. As newer chemistry mature and supporting infrastructure evolves, drones will become more autonomous, longer-lasting, and better suited to a wider array of complex tasks. In warfare, where success often hinges on speed, endurance, and technological edge, the right battery can be the difference between mission success and failure. For both civilian innovation and battlefield dominance, advancing drone battery technology remains not just a technical priority, but a strategic imperative.

CHAPTER 7

Drone Batteries Shared Technologies

4.4.1 Shared Electrochemical Foundations

Drone batteries—whether Lithium Polymer (LiPo), Lithium-Ion (Li-Ion), or the emerging Lithium Polymer Solid-State Batteries (Li-PSB)—share a fundamental electro-chemical foundation rooted in the movement of lithium ions between the anode and cathode during charge and discharge cycles. Despite the differences in structure, material composition, and performance characteristics, all three types operate on the same basic principle: the reversible intercalation and deintercalation of lithium ions through an electrolyte medium, enabling energy storage and delivery. This shared foundation provides a common platform for comparison, while variations in electrolyte composition, separator type, electrode materials, and packaging technologies result in differing performance, safety, and application suitability.

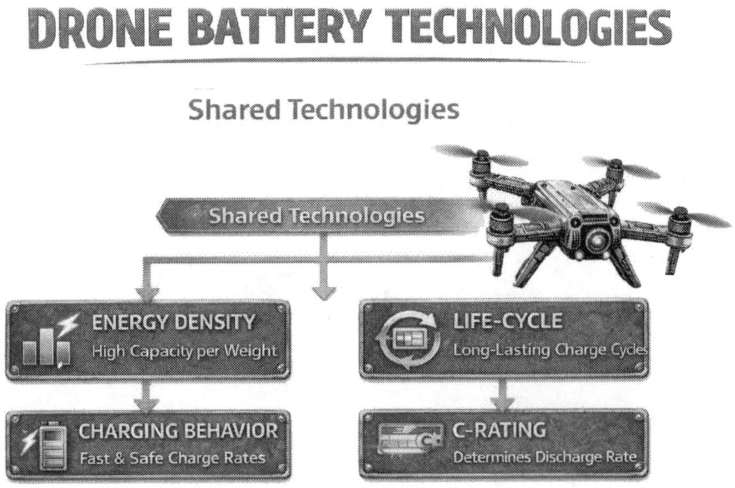

The essential electro-chemical process in all lithium-based batteries involves lithium ions migrating from the positive electrode (cathode) to the negative electrode (anode) during charging, and reversing direction during discharge. The anode typically comprises a carbon-based material, most commonly graphite, while the cathode consists of a lithium metal oxide compound such as lithium cobalt oxide ($LiCoO_2$), lithium iron phosphate ($LiFePO_4$), or lithium nickel manganese cobalt oxide ($LiNiMnCoO_2$). The electrolyte—whether liquid, gel, or solid—facilitates the ionic movement while acting as an electrical insulator between electrodes to prevent short circuits.

In Li-Ion batteries, the electrolyte is a liquid organic solvent containing lithium salts such as lithium hexafluorophosphate ($LiPF_6$). This type of electrolyte allows for high ionic conductivity and supports fast charge and discharge rates, making Li-Ion batteries highly efficient. They are known for their high energy density, relatively long cycle life, and stable voltage discharge profiles, which make them suitable for applications requiring sustained energy output over extended periods. Li-Ion batteries are widely used in fixed-wing drones, hybrid UAVs, and long-endurance applications where consistent power delivery is prioritised over burst performance. However, the flammable nature of their liquid electrolytes poses thermal and safety challenges, especially under physical stress, overcharging, or high-temperature conditions.

Lithium Polymer (LiPo) batteries build on the same electro-chemical principles but replace the rigid cylindrical or prismatic casing of Li-Ion cells with a flexible pouch format and use a gel-like polymer electrolyte instead of a pure liquid. This design choice improves energy-to-weight ratios and enables more flexible battery shapes, which is a significant advantage in drone design where space and weight are critical. LiPo batteries generally offer higher discharge rates than Li-Ion batteries, supporting rapid energy delivery for high-thrust applications like vertical take-off, racing drones, and agile rotary-wing UAVs used in tactical scenarios. Their C-ratings (a measure of discharge rate) are often substantially higher, enabling quick acceleration and responsive manoeuvring. However, this comes at the cost of reduced cycle life and increased sensitivity to physical damage and improper charging. LiPo cells can swell, overheat, or catch fire if mishandled, which presents a significant operational risk, particularly in rugged or combat conditions.

Emerging as a solution to the safety and performance limitations of both Li-Ion and LiPo chemistry is the Lithium Polymer Solid-State Battery (Li-PSB). This next-generation battery type retains the flexible pouch format of LiPo cells but replaces the liquid or gel electrolyte with a solid or semi-solid polymer-based electrolyte. This structural change offers multiple advantages. The solid electrolyte is non-flammable, reducing the risk of thermal runaway or combustion in case of impact, overheating, or short circuit. It also enables higher energy densities and better thermal stability, which are crucial for mission-critical UAV operations in extreme environments. Because the solid electrolyte does not leak or degrade as rapidly as liquid electrolytes, Li-PSBs also offer improved cycle life, making them more cost-effective and reliable over long-term use.

From an electro-chemical perspective, Li-PSBs still operate via lithium ion transport across the electrolyte during charge and discharge, but the ion mobility and interfacial characteristics differ significantly. One of the technical challenges in Li-PSB development is maintaining high ionic conductivity within the solid electrolyte and ensuring good interfacial contact between electrodes and the electrolyte material. Poor contact or high resistance at the interface can reduce battery performance, increase internal heating, and limit charging speeds. Advances in material science, such as the use of doped ceramic-polymer composites or engineered thin-film layers, are helping to overcome these issues, making Li-PSBs increasingly viable for commercial and military drone applications.

When comparing performance metrics, Li-Ion batteries typically offer energy densities in the range of 150–250 Wh/kg, with moderate discharge rates and long cycle life. They are ideal for endurance missions where weight efficiency and sustained power are essential. LiPo batteries offer slightly lower energy densities—usually 130–200 Wh/kg—but significantly higher discharge rates, often exceeding 25C or more, which supports rapid energy delivery and agility. Their flexible form factor is also highly valued in compact UAV designs. Li-PSBs are projected to exceed both Li-Ion and LiPo in energy density, potentially reaching 300–400 Wh/kg as the technology matures, with discharge rates comparable to LiPo batteries but with vastly improved safety and operational lifespan.

Another comparison point is temperature performance. Li-Ion and LiPo batteries both experience performance degradation in cold environments due to reduced electrolyte conductivity, leading to lower discharge capacity and slower charging rates. Li-PSBs, depending on the specific electrolyte formulation, can also be sensitive to temperature extremes, but their thermal stability and lower risk of expansion or rupture under heat give them an advantage in rugged and variable environmental conditions. This makes them especially appealing for military drones deployed in deserts, arctic zones, or high-altitude environments where reliability under temperature stress is paramount.

Charging behaviour is also a distinguishing factor. Li-Ion and LiPo batteries can typically be recharged within one to two hours using high-current chargers, but this process requires precise voltage control and thermal monitoring to avoid degradation or failure. Li-PSBs currently exhibit slower charging rates due to limitations in ion mobility across the solid-state medium. However, this drawback is being addressed through innovations in battery design and thermal regulation, and future iterations are expected to support faster, safer recharging without compromising cell integrity.

In terms of infrastructure and integration, Li-Ion and LiPo batteries benefit from decades of development, with mature supply chains, standardised connectors, and widely available charging equipment. Li-PSBs are still in the early stages of commercial adoption, meaning recharging infrastructure, battery management systems (BMS), and compatible drone platforms are not yet as widely available. Nevertheless, because PSBs can be designed to mimic the dimensions and electrical profiles of existing LiPo packs, retrofitting or integration is technically feasible, and it is likely that future drones will be developed with native support for PSB technology as it becomes more mainstream.

In conclusion, while LiPo, Li-Ion, and Li-PSB drone batteries all rely on the same fundamental electrochemical principles of lithium ion transport between electrodes, their structural differences and material compositions create distinct profiles in performance, safety, cost, and application. Li-Ion batteries are well-suited for long-duration, energy-efficient missions; LiPo batteries are favoured for high-thrust, high-responsiveness roles; and Li-PSBs are positioned to combine the best of both worlds—offering high energy density, fast discharge, and superior safety. As battery technologies continue to evolve, drones will benefit from longer flight times, increased payload capacities, and improved reliability, making the choice of battery not just a technical consideration but a strategic decision in the design and deployment of UAV systems.

4.4.2 Why Battery Energy Density Matters

Battery energy density plays a pivotal role in drone flight, acting as a key determinant of performance, range, and utility. In technical terms, energy density—measured in watt-hours per kilogram (Wh/kg)—refers to the amount of energy a battery can store relative to its weight. For drones, which are inherently constrained by payload capacity and aerodynamic efficiency, every gram counts. The higher the energy density of a drone's battery, the more energy it can store without significantly increasing its weight, directly translating to longer flight times, greater range, and improved mission versatility.

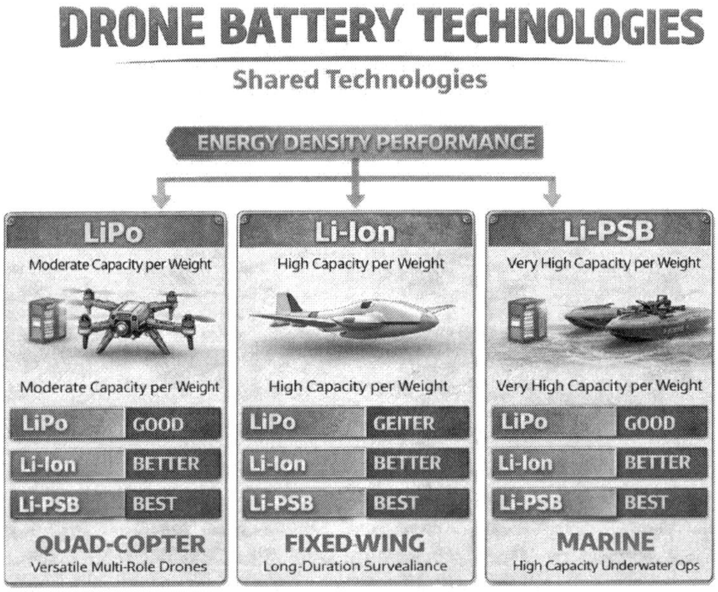

Battery energy density sits at the very core of modern drone capability, defining the practical limits of endurance, payload integration, and overall mission effectiveness. In unmanned systems, every gram allocated to energy storage must justify itself in performance return, making energy density—not simply battery size—the decisive factor in design efficiency. Higher energy density allows greater power storage without proportional increases in weight or volume, directly translating into longer flight times, expanded operational range, and the ability to support more advanced onboard systems. Whether for high-resolution imaging platforms, complex sensor arrays, communications relays, or mission-specific payloads, the capacity to store more energy per unit mass fundamentally reshapes what a drone can achieve within a single deployment cycle.

Emerging technologies, particularly solid-state lithium polymer and next-generation lithium-based chemistry, promise substantial improvements in energy density compared to traditional lithium-ion cells. If commercially viable at scale, these batteries could dramatically increase endurance across both rotary-wing and fixed-wing platforms without compromising aerodynamic efficiency. For multi-rotor drones, which are typically constrained by high power draw, improved energy density could double operational flight times. Fixed-wing systems, already optimised for glide efficiency, would benefit even more significantly, potentially extending airborne duration into multi-hour mission profiles. In practical terms, enhanced energy density reduces logistical burdens, minimises battery swap frequency, and expands mission flexibility across reconnaissance, monitoring, and long-range operations.

From a broader technological perspective, energy density is the foundational enabler of autonomy. Stable, high-density power systems support advanced onboard processing, artificial intelligence integration, and (ML), machine learning-driven data acquisition without prematurely draining reserves. As drones increasingly function as mobile sensor platforms collecting and analysing real-time data, consistent and abundant energy becomes critical to sustaining computational loads alongside propulsion demands. Ultimately, improvements in battery energy density are not incremental enhancements—they are transformative multipliers that will determine the next generation of endurance, efficiency, and operational sophistication in unmanned aerial systems.

4.4.3 Emerging Technologies, Advanced LiPo

Advanced lithium polymer batteries represent a promising frontier in energy storage, particularly for applications such as drones where weight, safety, and performance are critical. Though they are not yet widely available on the commercial market, their potential is undeniable: delivering significantly more energy in a smaller, lighter, and inherently safer form factor. At the heart of this advancement are gel polymer electrolytes (GPEs), which offer a unique hybrid solution by bridging the properties of liquid and solid-state systems. These semi-solid materials consist of a polymer framework that's swollen with a liquid electrolyte, resulting in a system that retains the mechanical stability of a solid while preserving the high ionic conductivity of a liquid.

One of the key innovations in GPE-based lithium polymer batteries is the incorporation of inorganic fillers such as silicon dioxide (SiO_2), titanium dioxide (TiO_2), aluminum oxide (Al_2O_3), zirconium dioxide (ZrO_2), and lithium lanthanum zirconium oxide (LLZO). These compounds are integrated into the polymer-salt matrix to enhance several critical properties. They improve lithium-ion mobility by introducing additional conductive pathways, and they boost both mechanical and thermal stability—making batteries more robust under stress or high temperatures. Additionally, the fillers help suppress crystallinity in the polymer structure, which increases the flexibility of the polymer chains and allows ions to move more freely. This improvement in segmental motion is vital for maintaining high conductivity, particularly at ambient conditions. Furthermore, these additives reduce interfacial resistance between the electrolyte and electrode, which is often a performance bottleneck in conventional battery architectures.

The performance of GPEs also depends significantly on the choice of polymer host. Materials such as poly (vinylidene fluoride-co-hexafluoropropylene) (PVDF-HFP), poly(ethylene oxide) (PEO), and poly(methyl methacrylate) (PMMA) are commonly employed due to their excellent compatibility with lithium salts and their ability to form stable gels. These polymers provide a balance of mechanical resilience and elector-performance, helping to stabilise the electrolyte structure while ensuring efficient ion transport.

In summary, advanced lithium polymer batteries using gel polymer electrolytes and inorganic fillers are shaping up to be game-changers for next-generation drone platforms. They promise longer flight durations, safer operation, and increased energy density—all without compromising weight or size. As research continues to refine these materials, their adoption could mark a turning point in the evolution of compact, high-performance power systems.

4.4.4 Battery Comparison, Drone Type, Form Factor

Solid-state LiPo batteries offer a range of advantages and trade-offs depending on the type of drone platform. For rotary-wing drones—such as quad-copters and vertical takeoff and landing (VTOL) systems—these batteries are particularly well-suited due to their ability to meet the high power demands of such platforms. Rotary drones require batteries that can deliver high current, respond rapidly to throttle changes, and maintain voltage stability under heavy loads. Solid-state LiPo batteries excel in these areas because of their low internal resistance and superior thermal stability, which allows them to provide consistent current without excessive voltage sag during rapid acceleration or sudden maneuvers. This makes them highly effective in operational scenarios such as urban reconnaissance, close-quarters strike missions, tactical swarm operations, and search and rescue missions in rugged or confined environments. With improved safety margins and endurance, drones powered by solid-state LiPo cells can loiter for longer periods, maneuver more aggressively, and support heavier payloads, including thermal imaging systems, jamming equipment, or small munitions.

For fixed-wing and hybrid drones, the benefits of solid-state LiPo batteries are slightly different. These platforms prioritise endurance and efficiency over rapid manoeuvrability. In this context, the increased energy density of solid-state LiPo batteries has become crucial, as it can potentially double or even triple a drone's operational range. This expanded range allows for deeper reconnaissance missions, greater persistence in intelligence-gathering, and longer loiter times without the need for constant GPS relay or battery swaps. Solid-state LiPo-powered fixed-wing drones could even replace traditional fuel-based UAVs in some applications, reducing the logistical burden of transporting fuel and simplifying field operations. The lightweight, compact nature of these batteries also supports more streamlined air-frame designs, enabling improved aerodynamics and further enhancing mission endurance.

When examining battery form factors and their impact on design, key differences emerge between Li-Ion and LiPo technologies. Li-Ion batteries typically come in cylindrical (such as 18650 or 21700) or prismatic formats, housed in rigid metal enclosures with built-in protection circuitry for over-voltage, under-voltage, and thermal control. These batteries are optimised for safety, long life cycles, and energy efficiency, making them ideal for consumer electronics, high-endurance fixed-wing UAVs, and hybrid drone systems where extended operation times are prioritised over rapid power delivery.

In contrast, LiPo batteries utilise a solid polymer or gel-based electrolyte contained within flexible, soft pouches. This form factor allows for custom shaping, which is advantageous when integrating batteries into irregular drone fuselages or compact builds. LiPo batteries are particularly favoured in RC and FPV racing drones, high-thrust rotary-wing platforms, and improvised or battlefield-modified UAVs where a high discharge rate (C-rating) is essential. Their flexibility, high power output, and low weight make them indispensable in applications where responsiveness and power density outweigh longevity and energy efficiency.

CHAPTER 8

(ICE) Internal Combustion Technologies and Hybrids

5.1.1 ICE Types in Drones

In military drone operations, ICE commonly refers to Internal Combustion Engine propulsion systems, which provide an alternative to fully electric power-trains. Unlike battery-dependent platforms, ICE-powered drones offer substantially greater range, increased payload capacity, and extended endurance—characteristics that are especially valuable during prolonged missions in contested or remote environments.

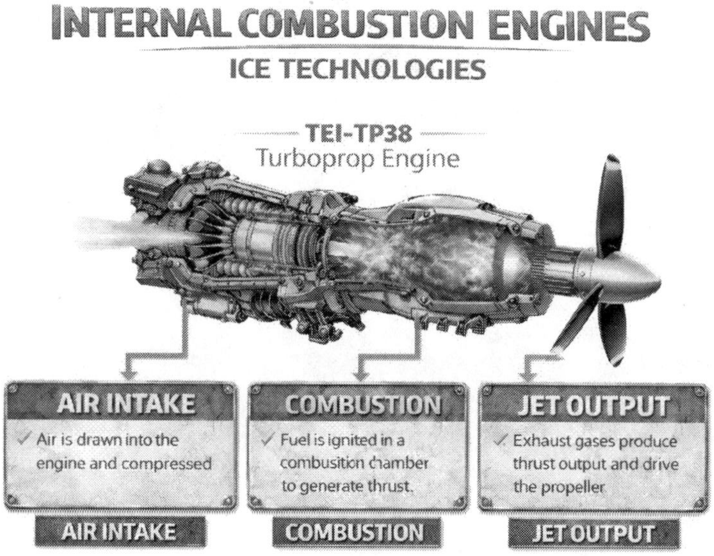

Within the evolving adaptation of hobby-derived drones for warfare, ICE propulsion has experienced steady—and in some sectors significant—advancement. As conflicts have extended over prolonged operational periods, lightweight polystyrene and wood-based hobby air-frames have increasingly been modified to integrate compact combustion engines. These materials remain attractive due to their low cost, rapid assembly, and ease of field repair, making them suitable for scalable and expendable deployment models. When paired with internal combustion propulsion, such platforms can achieve flight durations that surpass battery-powered equivalents, enabling deeper operational reach, extended loiter capability, and improved mission flexibility in sustained engagements.

This fusion of traditional model aircraft construction with modern propulsion systems reflects the adaptive, iterative character of contemporary drone development. Beyond improvised adaptations, ICE systems continue to hold a defined role across formal military drone architectures. Fixed-wing ICE platforms, often resembling small conventional aircraft, are engineered for endurance-intensive missions including long-range surveillance, reconnaissance, and precision strike operations. Their aerodynamic efficiency combined with fuel-based propulsion enables extended time-on-station performance. Systems such as the MQ-9 Reaper and the Bayraktar Akıncı demonstrate how combustion-powered UAVs can remain airborne for extended durations, supporting persistent intelligence and strike capabilities.

Rotary-wing ICE drones, while less common than electric variants, offer advantages in lift capacity and sustained flight under heavier loads. Their vertical take-off and landing capability provides operational access in confined or infrastructure-limited environments, though such configurations are typically deployed for specialised high-lift or endurance-focused missions. Hybrid ICE-electric propulsion systems have further expanded flexibility, combining combustion engines for cruise efficiency with electric motors for take-off, landing, or low-signature flight phases. This blended approach improves fuel economy while reducing acoustic and thermal exposure during sensitive operations. Additionally, loitering munitions equipped with combustion engines—sometimes termed kamikaze drones—leverage extended endurance to remain over target areas until engagement opportunities arise, offering adaptability in time-sensitive scenarios. Collectively, these ICE-powered configurations highlight the continued strategic relevance of combustion propulsion in modern drone warfare, particularly where endurance, payload capacity, and operational persistence remain decisive factors.

5.1.2 ICE Key Technologies and Advantages

A key advantage of internal combustion engine (ICE) drones in warfare lies in their superior fuel efficiency and extended operational range. By utilising energy-dense liquid fuels, these drones can undertake missions that stretch across hundreds or even thousands of kilometres—far surpassing the endurance of battery-powered counterparts. Their robust thrust capabilities also translate to a significantly greater payload capacity, allowing them to carry heavier or more sophisticated equipment such as high-resolution sensors, secure communication relays, or weapon systems. This makes them especially well-suited for complex, multi-role operations. Additionally, their ability to remain airborne for prolonged periods enhances their value in intelligence, surveillance, and reconnaissance (ISR) missions, as well as in electronic warfare where persistent presence is essential. Perhaps one of the most pragmatic benefits is their capacity for rapid refuelling. Unlike electric drones, which may require extended downtime to recharge, ICE drones can be quickly topped up and sent back into the field, maximising operational tempo and strategic flexibility in active conflict zones.

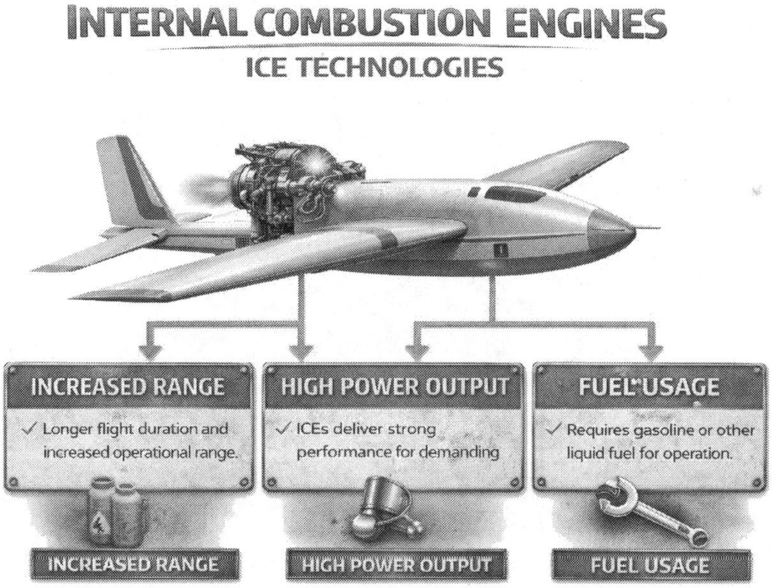

5.1.3 ICE Limitations and Challenges

While internal combustion engine (ICE) drones offer significant advantages in terms of range, payload capacity, and operational endurance, their deployment in drone warfare is not without considerable limitations and challenges. One of the most pressing issues is detectability. ICE drones produce considerable noise and thermal signatures due to their combustion process, making them easier to track with acoustic sensors, infrared targeting systems, or radar. This heightened visibility can compromise stealth-dependent missions and limit their effectiveness in contested or high-threat environments where undetected operation is crucial.

Moreover, ICE drones are mechanically more complex than their electric counterparts. Their engines consist of numerous moving parts that are subject to wear and tear, increasing the potential for mechanical failures in the field. This not only reduces reliability but also demands regular and often specialised maintenance, which can be logistically challenging in austere or rapidly shifting combat zones. In contrast, electric drones —thanks to their simpler propulsion systems—tend to be more robust and easier to service or replace on short notice.

Fuel logistics present another layer of complexity. While ICE drones benefit from using energy-dense fuels, they are reliant on consistent and secure fuel supply chains. In wartime scenarios where supply lines are vulnerable to disruption, maintaining access to aviation fuel or other specific combustibles may become problematic. This dependency can constrain long-term operations or deployments in remote regions without established refuelling infrastructure.

Environmental considerations also pose significant hurdles. ICE drones emit greenhouse gases and other pollutants, which not only have ecological impacts but can also affect drone performance in specific operational settings, such as high-altitude missions where engine efficiency drops due to reduced oxygen levels. Furthermore, the vibrations and heat generated by ICE propulsion can interfere with the sensitive electronics and payloads commonly used in modern drone warfare, such as high-precision cameras or communications systems.

Finally, the growing advancement in counter-UAV technology means ICE drones may become increasingly vulnerable. Their audible engine noise makes them more susceptible to acoustic detection, and their larger radar cross-section—due to size and fuel systems—can render them easier targets for anti-air systems.

In summary, while ICE drones remain a critical tool in modern warfare due to their endurance and power, they face a range of operational, logistical, and tactical limitations. Balancing these trade-offs is essential for mission planning and may explain the increasing trend toward hybrid or fully electric systems in drone warfare evolution.

5.1.4 ICE Innovations in Technology

Innovations in internal combustion engine (ICE) drone technology have significantly evolved in recent years, driven by the increasing demand for longer-range, high-endurance unmanned systems in military, industrial, and surveillance applications. One of the most impactful advancements lies in the miniaturisation and optimisation of multi-fuel engines. Modern ICE drones are being equipped with lightweight, high-efficiency engines capable of running on a variety of fuels, including heavy fuels like JP-8, which are widely used in military operations for logistical simplicity. These engines are designed for improved thermal efficiency, reduced vibration, and greater reliability, allowing for extended flight durations without compromising structural integrity.

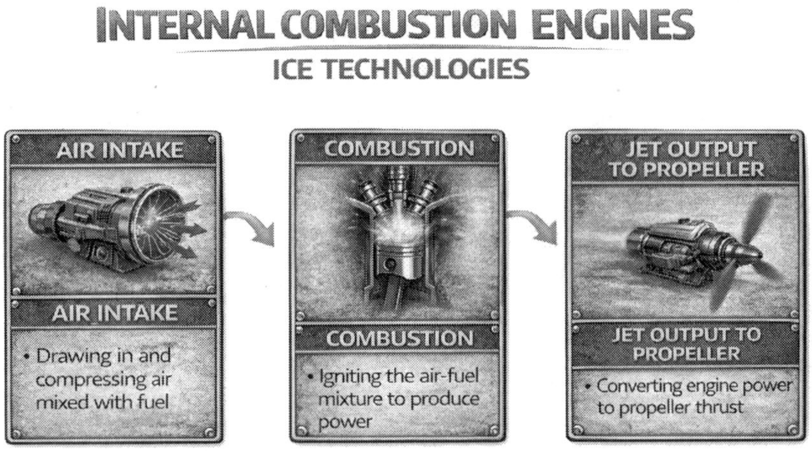

Another key development is the integration of autonomous engine management systems. These onboard processors continuously adjust fuel flow, ignition timing, and air intake based on flight conditions, altitude, and mission parameters. This kind of adaptive engine control not only improves performance under varying loads and climates but also extends engine lifespan by minimising wear and thermal stress. Coupled with advancements in lightweight materials—such as carbon composites and advanced alloys—ICE drones are now lighter and more resilient, supporting both increased payload capacities and enhanced agility in the air.

Thermal signature reduction has also become a focal point in recent ICE drone innovations, particularly for military platforms. Engineers are employing advanced exhaust management systems, including redirected airflow designs, heat-dissipating coatings, and insulated shrouds to lower infrared detect-ability. This is crucial for stealth operations, as it mitigates the vulnerability of drones to heat-seeking technologies and makes them more survivable in contested airspace.

Hybrid propulsion systems represent another trans-formative step forward. By combining ICEs with electric motors, drones can switch between power sources as needed—using electric propulsion for silent, low-speed operations and ICEs for cruise and endurance. This not only reduces acoustic signatures but also dramatically increases flight range and fuel economy. In some cases, hybrid systems are integrated with regenerative technologies, where rotors or propellers can recharge onboard batteries during descent or glide phases, enhancing overall energy efficiency.

Additionally, maintenance innovations such as predictive diagnostics are beginning to play a role in field operations. These systems use onboard sensors and machine learning algorithms to detect early signs of wear, misalignment, or engine imbalance, helping operators perform proactive maintenance and avoid mission-critical failures.

Collectively, these innovations are transforming ICE drones from fuel-hungry workhorses into smart, agile, and stealthy aerial assets, ensuring their continued relevance in modern and future conflict zones, as well as in long-endurance civilian missions where electric solutions are not yet feasible.

In modern conflicts like those in Ukraine and the Middle East, ICE-powered drones have proven indispensable for both state and non-state actors. Their ability to deliver precision strikes, conduct deep reconnaissance, and operate in GPS or comms-denied environments makes them a cornerstone of contemporary drone warfare.

5.1.5 ICE Technology, Use in Hobby Kits

Innovation in internal combustion engine (ICE) drone technology has not only reshaped civilian applications but is increasingly influencing modern warfare in relation to hobby drones, especially in asymmetric conflict scenarios. As warfare becomes more technologically dispersed and multi-domain, militaries and non-state actors are exploiting both cutting-edge and grassroots innovations to build highly capable unmanned systems. One of the most notable trends is the convergence of advanced ICE systems, lightweight composite materials, and the accessibility of hobby-grade drone kits—all of which combine to create low-cost, high-impact aerial platforms with battlefield utility.

Modern ICE hobby drones benefit from purpose-built combustion engines that are not only smaller and lighter but are capable of operating on a range of fuels, including aviation-grade kerosene (like JP-8), gasoline, or even diesel blends. These engines offer greater thrust and energy density than electric counterparts, making them ideal for missions requiring extended flight durations, heavier payloads, or operations in regions where battery charging infrastructure is impractical. Innovations such as electronically controlled fuel injection, optimised turbocharging, and thermal efficiency improvements have made these engines more reliable and adaptable to changing environmental conditions. In military settings, this translates to tactical advantages such as longer surveillance missions, increased loitering time over hostile territory, and the ability to carry precision munitions or advanced sensors without compromising flight time.

The construction materials used in ICE drones have also undergone a significant transformation. Hobbyist and defense contractors are turning to lightweight composites such as carbon fibre-reinforced polymers (CFRPs), fibreglass, and high-tensile epoxy resins to construct air-frames that are not only durable but also radar-resistant and thermally stable. These materials provide high strength-to-weight ratios, which is crucial for increasing flight efficiency and payload capacity. Additionally, components like rotors, exhaust shrouds, and engine mounts are being fabricated from heat-resistant alloys and ceramics to withstand the high operating temperatures typical of combustion engines. The use of such materials is especially beneficial for drones designed for harsh environments or high-altitude reconnaissance, where thermal gradients and structural integrity are critical.

A particularly disruptive development in contemporary warfare is the militarization of hobbyist drone platforms. Commercially available hobby kits—originally designed for enthusiasts and recreational users—are being co-opted and modified for tactical use. These kits often include customisable light weight wooden frames, modular engine bays, and open-source flight controllers, making them ripe for adaptation. In recent conflicts, including in Ukraine and parts of the Middle East, we've seen these so-called "garage-built" drones outfitted with ICE engines for longer range and payloads as varied as surveillance pods, jamming equipment, and improvised explosive devices. In many cases, these platforms are fitted with low-cost GPS navigation, first-person-view (FPV) systems, and even rudimentary autonomous flight scripts, dramatically expanding their strategic potential.

The appeal of ICE-driven hobby drones in warfare lies in their cost-effectiveness and operational flexibility. A drone assembled from off-the-shelf parts, with a small ICE power-plant, can be launched from remote locations, fly for hours, and carry out missions that traditionally required manned aircraft or million-dollar UAV systems. Their expendable nature also adds to their utility: forces can use them in swarms, run decoy missions, or deploy them in GPS-denied environments with minimal financial risk. Moreover, the rise of 3D printing has made it possible to fabricate replacement parts or payload compartments on-site, further reinforcing the modular and scalable nature of these platforms.

However, the democratisation of drone warfare—where increasingly affordable and accessible technologies allow state and non-state actors alike to deploy unmanned systems—introduces a complex mix of tactical opportunities and significant security challenges. On one hand, the resourceful use of internal combustion engine (ICE)-powered drones provides notable advantages in terms of endurance, payload capacity, and operational range. These drones are particularly valuable in extended missions where battery-powered systems may fall short, such as long-range reconnaissance, persistent surveillance, or kamikaze-style strikes. Their ability to be refuelled rapidly and operate in areas with limited charging infrastructure makes them highly adaptable in austere or contested environments. Moreover, the relative ease of sourcing ICE components from commercial markets enables irregular forces and militias to assemble capable systems without access to high-end manufacturing or military supply chains.

Yet, this same accessibility poses a threat to global security. The proliferation of such platforms lowers the barrier for armed groups, terrorist cells, or rogue states to conduct asymmetric attacks, disrupt infrastructure, or challenge conventional military forces. Without stringent export controls, technical safeguards, or effective counter-drone measures, ICE-powered drones could become instruments of destabilisation, especially when combined with improvised explosives or autonomous targeting software. Thus, while democratised drone warfare empowers more actors on the battlefield, it also demands a coordinated international response to manage its risks.

5.1.6 ICE Summary, Technology

In conclusion, the integration of internal combustion engine (ICE) technology into hobby drones for warfare represents a significant shift in how low-cost, off-the-shelf systems are adapted for military use. While electric drones remain dominant in many tactical roles due to their low noise profile and ease of use, ICE-powered drones are increasingly favoured in situations requiring greater endurance, heavier payloads, and operational flexibility. Hobby drones equipped with ICE engines can sustain flight for longer periods without the limitations of battery life, making them ideal for missions involving extended surveillance, long-distance reconnaissance, or even improvised strike operations over hundreds of Km's. Their ability to carry heavier loads compared to typical battery-powered counterparts also makes them suitable for transporting small munitions, communications gear, or logistical supplies in areas that are otherwise difficult to reach.

The key appeal of ICE technology in this context lies in its accessibility and adaptability. Civilian hobbyists and DIY engineers can often retrofit or modify model aircraft platforms to include small internal combustion engines, effectively creating a powerful and inexpensive UAV platform with minimal resources. This has made ICE drones attractive not only to conventional forces but also to insurgent groups and irregular actors operating outside the constraints of traditional defence procurement channels. The capacity to procure fuel easily and operate independently of sophisticated charging infrastructure further enhances the value of ICE-powered systems in austere or contested environments.

However, ICE-powered drones require more complex maintenance and technical skill to operate, particularly in harsh conditions where mechanical reliability is critical. Fuel logistics can also become a constraint, especially in remote or besieged areas where resupply is limited.

Despite these limitations, the continued use and evolution of ICE hobby drone technology in warfare underscores the changing nature of conflict and the increasing role of civilian-derived systems in modern battlefields. As drone warfare becomes more decentralised and democratised, ICE-powered drones offer a compelling blend of affordability, capability, and adaptability that appeals to a wide spectrum of actors. Their presence in modern conflicts—from Ukraine to the Middle East—signals not just a technological adaptation, but a strategic evolution in how air-power is conceptualised at the small-unit and tactical level. As such, understanding and countering the use of ICE hobby drones will remain a critical focus for military planners and defence analysts in the years ahead.

5.2.1 Gas Turbines and Jet Engines

Jet engines and gas turbine propulsion systems are typically reserved for very large, high-speed military drones such as target drones, cruise missiles, and long-range surveillance UAVs, rarely seen in hobby drones. These drones are designed for missions that demand extremely high thrust and sustained high-speed flight, often at altitudes and distances far beyond the capabilities of traditional electrically powered drones. The advantage of such propulsion systems lies in their ability to generate immense amounts of power, allowing unmanned aerial vehicles to cruise at supersonic or high subsonic speeds while maintaining stability and responsiveness in complex airspace.

This makes them ideal for penetrating deep into hostile territory, evading radar detection, or delivering payloads with great precision over long distances. For instance, cruise missiles powered by turbojet or turbofan engines are capable of traversing thousands of kilometres with pinpoint accuracy, guided by onboard navigation systems. Similarly, high-speed target drones powered by small turbojet engines are used by military for training purposes, simulating the flight profiles of enemy aircraft or missiles to test the effectiveness of defensive systems. Despite their performance advantages, these propulsion systems come with significant drawbacks that limit their applicability beyond military contexts.

They are prohibitively expensive to produce and maintain, consume large quantities of fuel, and generate considerable noise and heat, making them impractical for covert or environmentally conscious operations. Additionally, their complex engineering requires highly specialised materials and precise manufacturing, which adds to the overall cost and limits mass accessibility. These factors collectively make jet-powered UAVs unsuitable for use in the hobby drone space, where affordability, ease of maintenance, and noise reduction are typically higher priorities. In contrast, hobby drones often rely on electric motors due to their lower cost, minimal maintenance requirements, and quieter operation, which are more appropriate for recreational flying, aerial photography, or small-scale research.

The sheer speed and altitude capabilities of jet-propelled drones also make them difficult to control without advanced avionics and flight control systems, placing them well beyond the skill-set and resources of most civilian users. Moreover, regulatory and safety constraints further restrict the use of such high-performance drones in civilian airspace, with stringent requirements on airworthiness, flight paths, and risk mitigation.

Thus, while jet and turbine propulsion technologies represent the pinnacle of drone performance in military applications, they remain a niche and highly specialised domain, rarely seen in the broader landscape of hobbyist or commercial drone activity.

5.3.1 Hybrid Power Systems

Hybrid power systems in UAVs combine internal combustion engines (ICE), fuel cells, or solar technologies with batteries to optimise endurance, efficiency, and performance. One common configuration is the ICE-electric hybrid, where a combustion engine is used to either recharge onboard batteries or provide supplementary power during cruise phases, thereby extending flight time while reducing reliance on large battery packs alone. This setup is particularly effective in medium to large UAVs that require sustained flight with periodic bursts of high power.

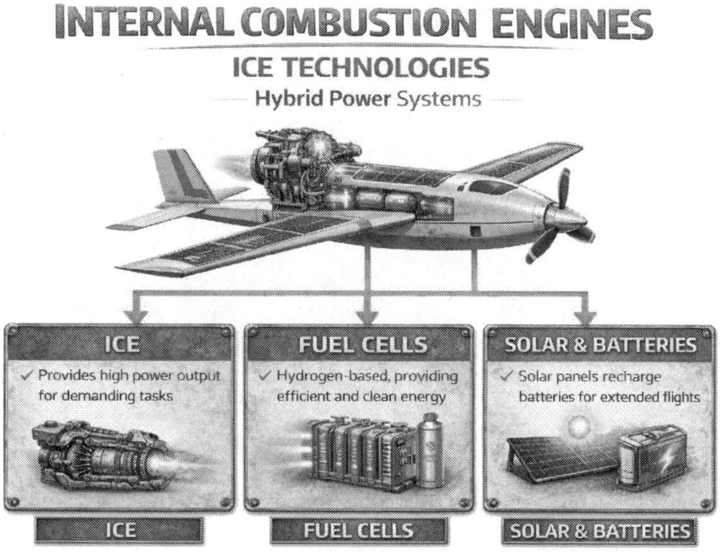

Another increasingly popular approach is the integration of hydrogen fuel cells with batteries. In this system, the fuel cell supplies continuous, steady power for general flight operations, while the battery kicks in to handle peak power demands such as takeoff, rapid ascent, or manoeuvring under load. This dual-source setup offers high energy density and cleaner emissions, making it attractive for longer-duration, environmentally conscious missions.

A third configuration involves combining solar panels with batteries, particularly in high-altitude, long-endurance (HALE) drones like the Zephyr. These drones use solar energy collected during daylight hours to continuously recharge their batteries, enabling them to stay airborne for weeks or even months at a time without the need for re-fueling or landing. Solar-electric hybrids are especially suited for surveillance, scientific observation, and communication relay missions where persistent flight at high altitudes is critical. By leveraging these hybrid systems, UAV designers can balance the strengths of each power source—whether it's the high energy density of fuel, the consistent output of fuel cells, or the renewable nature of solar energy—with the flexibility and responsiveness of modern battery technology.

5.3.2 Hybrid ICE and Electric

The use of internal combustion engine (ICE) and electric hybrid power systems in drones represents a strategic evolution in unmanned aerial vehicle (UAV) technology, aimed at enhancing flight endurance, payload capacity, and operational flexibility. This hybrid configuration combines the benefits of traditional fuel-driven engines with the precision and efficiency of electric propulsion. In a typical ICE-electric hybrid setup, the combustion engine is used either as the primary source of propulsion or more commonly as a generator to charge onboard batteries, which in turn power electric motors that drive the rotors or propellers. This approach allows drones to operate for longer durations than battery-only designs, while also reducing the overall weight burden associated with carrying multiple or larger battery packs. The hybrid system can switch between modes or operate simultaneously depending on flight demands—during high-power requirements like takeoff and climbing, both the engine and electric motors might operate in tandem, whereas during cruise or descent, the system may rely more on the electric component to conserve fuel.

This type of power-train is particularly beneficial for medium to large UAVs used in applications such as long-range surveillance, mapping, search and rescue, or cargo delivery, where flight duration and payload capacity are critical. The ability to recharge batteries in-flight means that the drone can sustain operations for many hours without requiring external charging infrastructure, a major advantage in remote or austere environments. Furthermore, by utilising electric motors during lower-power or sensitive phases of flight, such as hovering or flying over populated areas, the drone can reduce noise emissions and improve stealth characteristics—an important feature for military or law enforcement use. From a technical standpoint, hybrid systems also introduce redundancy; if one component of the power-train fails, the other can often compensate, improving reliability and safety.

However, ICE-electric hybrids come with trade-offs. They are mechanically more complex than purely electric systems, requiring careful integration of engine, generator, power management electronics, and electric propulsion units. This complexity can lead to increased maintenance demands and higher initial costs. Fuel logistics and engine servicing also become factors in operational planning. Despite these challenges, the hybrid approach is gaining traction due to its ability to bridge the gap between short-range electric drones and long-endurance fuel-powered aircraft. Innovations in lightweight engines, improved battery chemistry, and more efficient power management systems continue to improve the viability and performance of hybrid drones. As technology advances, ICE-electric hybrid UAVs are expected to play an increasingly important role in both commercial and defence sectors, offering a flexible and robust solution for a wide range of aerial missions.

5.3.3 Hybrid Fuel Cell, Solar and Battery

The combination of hydrogen fuel cells, batteries and/or solar cells in drone propulsion systems presents a highly promising solution for achieving long-endurance, environmentally friendly, and reliable flight operations. This hybrid power approach leverages the unique strengths of both energy sources: hydrogen fuel cells provide a steady and continuous supply of power, while batteries are employed to manage peak loads and high-demand situations such as takeoff, rapid ascent, or complex manoeuvring. In a typical configuration, the hydrogen fuel cell functions as the primary energy provider, steadily generating electricity through an electro-chemical reaction that combines hydrogen with oxygen, emitting only water vapor as a byproduct. This makes it an exceptionally clean and quiet energy source, particularly valuable for use in urban, ecological, or surveillance environments where low emissions and reduced noise signatures are essential. Batteries, usually lithium-based, are integrated into the system to handle power spikes that exceed the constant output capacity of the fuel cell, ensuring that the drone can perform high-energy tasks without compromising performance or efficiency.

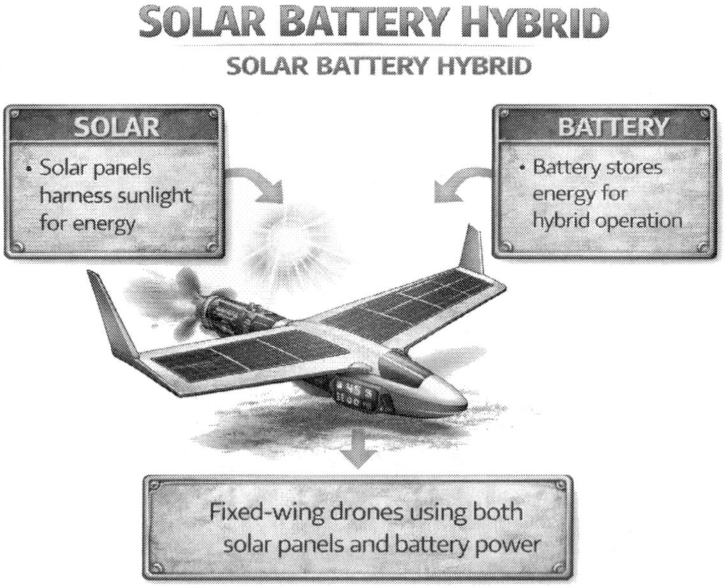

This dual-power strategy—integrating hydrogen fuel cells with batteries and, in some configurations, supplementary solar systems—has demonstrated particular effectiveness in medium to large UAVs designed for endurance and long-range mission profiles where sustained flight is critical and refuelling or recharging opportunities are scarce. Hydrogen's exceptional energy density by weight enables substantially longer airborne durations compared to battery-only systems, often extending operations to several hours while maintaining a relatively lightweight structural footprint. In many hybrid architectures, the fuel cell actively replenishes the onboard battery during flight, ensuring that electrical reserves remain available for peak-demand phases such as rapid ascent, evasive manoeuvring, or intensive sensor operation. This continuous energy loop enhances endurance, supports layered mission phases, and reduces dependence on fixed ground infrastructure.

Within the adaptable and innovation-driven environment of hobby-derived drone development, interest in hybrid propulsion systems has steadily expanded. Modular air-frames and open-architecture designs have allowed designers and engineers to experiment with integrating fuel cells alongside traditional electric systems, pushing operational endurance beyond the inherent limitations of standalone battery platforms. This iterative approach reflects the broader evolution of UAV propulsion strategies as developers seek greater range, reliability, and mission flexibility.

From an operational perspective, hydrogen-electric hybrids also present advantages in sustainability and logistical resilience. Hydrogen can be generated from renewable sources and stored in high-pressure containment systems, aligning propulsion development with global transitions toward cleaner energy technologies. Compared to internal combustion engines, fuel cells contain fewer moving parts, reducing mechanical wear, vibration, and maintenance demands over extended deployment cycles. However, practical challenges remain, including safe hydrogen storage, production economics, transport considerations, and the limited availability of refuelling infrastructure. Continued advancements in lightweight storage materials, compact fuel cell engineering, and green hydrogen production methods are progressively mitigating these constraints. As the technology matures, hybrid fuel cell–battery configurations are poised to play a defining role in advanced UAV propulsion, offering a refined balance of endurance, efficiency, and operational adaptability.

5.3.4 Hybrid Alternative Fuels (Experimental)

Hybrid alternative and experimental fuels are rarely seen in hobby drones, being within the confines of experimental use. However, the development and trial of experimental fuels for drones is an emerging field driven by the need for greater flight endurance, higher energy density, environmental sustainability, and adaptability to diverse operational contexts. Traditional drone propulsion has relied heavily on lithium-based batteries or small-scale internal combustion engines using gasoline or aviation fuels. However, these conventional energy sources present limitations in terms of flight time, weight, fuel availability, and emissions. As a result, researchers and manufacturers are now exploring alternative fuel options—ranging from bio-fuels and synthetic fuels to advanced hydrogen carriers and metal-air chemistry—to push the boundaries of drone performance and sustainability.

One promising area of development is bio-fuels, which are derived from renewable biological sources such as algae, plant oils, or agricultural waste. These fuels can be engineered to resemble conventional aviation fuels but with a significantly lower carbon footprint. Bio-fuels have the potential to be used in modified combustion engines or turbine-powered UAVs, offering a cleaner and potentially renewable alternative without requiring an entirely new propulsion architecture. Similarly, synthetic fuels produced through processes like Fischer-Tropsch synthesis or power-to-liquid technology use captured carbon dioxide and hydrogen to create liquid fuels, essentially recycling atmospheric carbon into usable energy. These synthetic fuels can also be adapted to current engine designs and are being tested in military and commercial drone platforms for their performance and logistical benefits.

Hydrogen-based experimental fuels are also being investigated beyond traditional gas forms. Alternatives like liquid hydrogen or hydrogen stored in solid-state materials offer improved energy density and safer handling characteristics. Metal hydrides and chemical hydrogen storage systems are being explored to enable longer storage durations and reduce the infrastructure burden associated with gaseous hydrogen tanks. Another groundbreaking concept in experimental fuels involves metal-air batteries, particularly aluminium-air and zinc-air chemistry, which promise extremely high energy density. These batteries use metal as fuel, reacting with oxygen from the air to generate electricity, making them lightweight and theoretically capable of supporting ultra-long-range missions. While promising, metal-air systems face significant challenges related to recharging and consistent power output.

Additionally, hybrid experimental solutions that combine multiple emerging fuel technologies are under investigation to optimise performance across a range of mission profiles. Despite their potential, most experimental fuels are still in early stages of testing, often within controlled environments or limited field trials. Challenges such as fuel stability, safety, environmental conditions, and compatibility with existing propulsion systems must be resolved before these fuels can be adopted at scale. Nonetheless, with increasing global focus on carbon neutrality, energy independence, and performance innovation, the field of experimental drone fuels is poised for rapid advancement, potentially revolutionising how UAVs are powered in both civilian and military domains.

Experimental fuels, while showing great potential for enhancing drone capabilities in terms of endurance, power output, and environmental impact, are understandably not widely adopted in hobby drones used in warfare due to their developmental status, limited availability, and practical constraints. These advanced fuels—such as hydrogen in solid or liquid form, synthetic fuels, bio-fuels, and metal-air battery chemistry—are typically confined to controlled research environments or high-budget military programs. Their complex handling requirements, high cost, and dependence on specialised storage and delivery systems make them largely inaccessible for informal or improvised combat drone applications, which often rely on off-the-shelf consumer technology. Hobby drones repurposed for conflict zones are usually selected for their affordability, ease of modification, and logistical simplicity, relying primarily on lithium-polymer (LiPo) batteries or small gasoline engines. These conventional power sources are readily available, well-understood, and easy to replace or recharge in the field—an essential factor in asymmetric or rapidly evolving combat environments. In contrast, experimental fuels require rigorous safety protocols, temperature control, and often sophisticated infrastructure, making them unsuitable for the ad hoc, low-cost warfare tactics commonly associated with hobbyist drones. Until these fuels become more mature, scalable, and user-friendly, their use in battlefield hobby drone applications is likely to remain extremely limited.

CHAPTER 9

UAV's Communication, Types and Technologies

6.1.1 UAV Communication, Introduction

The use of hobby drones in warfare has introduced a disruptive dynamic on the modern battlefield, driven largely by their affordability, availability, and adaptability. While initially designed for recreational flying, photography, and racing, hobbyist drones—particularly quad-copters and fixed-wing models—have been repurposed with lethal efficiency in asymmetric and conventional conflicts alike. From Ukraine to the Middle East, these small unmanned aerial vehicles (UAVs) have been retrofitted with improvised explosive devices, thermal cameras, and electronic warfare payloads, making them potent force multipliers. A critical enabler of their battlefield utility is the communication technology that underpins UAV operation. Without reliable communication links, drones cannot perform reconnaissance, deliver payloads, or conduct coordinated swarm attacks effectively.

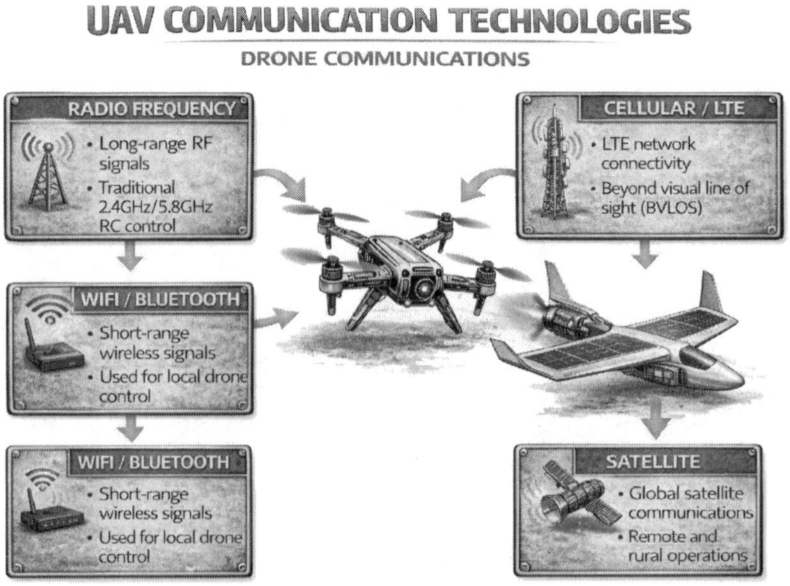

UAV communication systems are typically divided into three primary categories: line-of-sight (LOS), beyond-line-of-sight (BLOS), and autonomous or semi-autonomous systems. Hobby drones have almost exclusively operate on line-of-sight or short-range communications using radio frequency (RF) signals in the 2.4 GHz and 5.8 GHz bands. These frequencies are ideal for low-latency, high-resolution video transmission and control signals, making them well-suited for tactical operations in dense urban environments or close-range engagements. Many hobby-grade systems use analogue video transmitters (VTX) paired with goggles or monitors to provide first-person-view (FPV) capabilities, allowing the pilot to "see" through the drone's onboard camera in real time. While effective for short-range missions, analogue systems are vulnerable to interference and jamming. However, driven by the need for greater operational range, these types of hobby drones and UAVs have even been fitted with increasingly sophisticated communications systems including long range tethered and untethered fibre optic connections.

Digital FPV systems, pioneered by companies like DJI, have improved image quality and range, often using encrypted signals and stronger antennas to resist basic countermeasures. Some advanced systems integrate dual-band transmission and frequency-hopping spread spectrum (FHSS) techniques to reduce the risk of signal disruption. These technologies allow hobby drones to remain agile in contested electromagnetic environments, a feature increasingly exploited in wartime settings. Pilots can now execute more complex flight patterns, conduct detailed surveillance, or perform strike missions with greater precision, often while staying hidden from view.

For longer-range hobby drones or converted fixed-wing platforms, communication technologies may include telemetry radio modules that operate in the 900 MHz or 433 MHz bands. These systems enable real-time data exchange, including GPS coordinates, altitude, battery status, and other flight metrics, which are critical for mission planning and execution. Paired with software such as BETAFlight, Mission Planner or QGroundControl, operators can establish way-points, no-fly zones, and automatic return-to-home protocols. When combined with GPS, these drones can fly pre-programmed missions semi-autonomously, reducing the need for constant manual control and allowing for strategic deployments such as kamikaze strikes or perimeter over-watch.

Beyond traditional RF communication, Wi-Fi and 4G/5G cellular networks are increasingly being leveraged in hobby drone warfare, particularly in areas with reliable infrastructure. These methods allow drones to be controlled via smartphones or tablets from significant distances. Cellular connectivity opens up a host of possibilities, including live-streaming video to remote command posts and real-time data sharing across distributed units. In urban combat zones where line-of-sight might be obstructed, leveraging mobile networks can maintain UAV operability where standard RF would fail. However, reliance on commercial infrastructure introduces vulnerabilities, such as susceptibility to network outages or cyber attacks.

Autonomous navigation systems are another key evolution in UAV communication technology. By integrating inertial measurement units (IMUs), magnetometers, and barometric pressure sensors with GPS data, drones can navigate complex environments with minimal human input. Some systems use visual-inertial odometry or optical flow sensors to maintain stable flight in GPS-denied environments, which is particularly valuable in indoor or underground operations. These features not only make drones harder to detect but also harder to disrupt, since they do not rely on continuous external signals.

In summary, hobby drones have become formidable assets in modern warfare due to rapid advancements in communication technologies. From basic analogue FPV systems to sophisticated digital telemetry, fibre optics and autonomous navigation, these UAVs can perform a wide array of combat and support roles. The ability to communicate securely and effectively—whether through RF, cellular networks, or onboard computing—greatly enhances their tactical value. As these technologies continue to evolve and proliferate, so too will the impact of hobby drones on the battlefield, necessitating new countermeasures, doctrines, and strategic thinking across militaries worldwide.

6.1.2 UAV Radio Frequency (RF) Signal Control

Unmanned Aerial Vehicles (UAVs) have historically relied heavily on radio frequency (RF) signal control for communication, navigation, and operation. The effectiveness and versatility of UAVs in both civilian and military contexts are directly influenced by the quality and type of RF technologies they use. From hobby drones to advanced tactical UAVs, RF communication systems form the backbone of control, telemetry, and payload management. Understanding the formats, frequencies, and associated technologies is crucial to grasping how these aerial platforms function and how their vulnerabilities can be addressed or exploited.

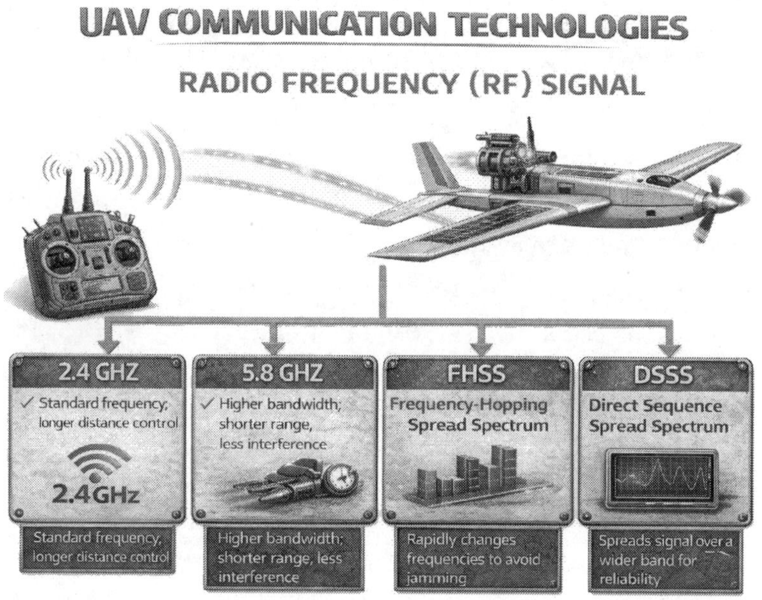

Most consumer and hobby-grade drones operate within the 2.4 GHz and 5.8 GHz ISM (Industrial, Scientific, and Medical) bands. These unlicensed frequencies are widely used due to their balance between range, data capacity, and hardware availability. The 2.4 GHz band offers relatively longer range and better penetration through obstacles such as buildings or vegetation, making it suitable for remote control (RC) links. The 5.8 GHz band, in contrast, is typically used for first-person view (FPV) video streaming due to its higher bandwidth, which allows for high-definition, low-latency video transmission. However, it is more susceptible to signal degradation over distance and through obstructions. Many drones use a dual-band system, where 2.4 GHz is dedicated to control signals and 5.8 GHz to live video feeds, ensuring real-time feedback while maintaining consistent operator control.

For enhanced communication reliability, many modern UAV systems use Frequency-Hopping Spread Spectrum (FHSS) or Direct Sequence Spread Spectrum (DSSS) technologies. These methods reduce the likelihood of signal jamming and interference by rapidly changing frequencies or spreading the signal across a wider bandwidth. FHSS, in particular, is popular in military and secure civilian applications, as it allows the control system to hop across multiple frequencies in a pseudo-random sequence known only to the transmitter and receiver. This not only makes signal interception and jamming more difficult but also enables better performance in RF-crowded environments.

Professional and long-range drones often expand beyond the ISM bands into less congested frequencies such as 433 MHz, 900 MHz, and 1.2 GHz, depending on regional regulations. These lower-frequency bands provide improved range and penetration capabilities, especially in environments with physical obstructions. The 433 MHz band is popular for telemetry data transmission in open-source drone systems, offering long-distance control at the cost of bandwidth and data speed. Likewise, the 900 MHz band—used extensively in North America—is ideal for beyond-visual-line-of-sight (BVLOS) operations, enabling data links that maintain stability even when the drone is several kilometres away from the operator.

More advanced hobby kit based UAVs and some industrial drones utilise custom RF protocols and encrypted communication formats. These are often proprietary and operate in reserved or classified frequency bands to avoid detection and ensure data integrity. Encrypted RF signals, sometimes using AES-256 or military-grade encryption standards, prevent adversaries from hijacking the UAV or intercepting sensitive data. Integration with satellite communications (SATCOM) allows for global coverage, especially when using C-band, Ku-band, or X-band frequencies. These systems, while more expensive and complex, are essential for strategic UAVs like the MQ-9 Reaper or RQ-4 Global Hawk, which operate over vast distances and require constant, secure command and control.

In recent years, there has been growing interest in 4G and 5G integration for UAV communication. Using mobile networks, drones can connect to control stations via cellular towers, bypassing traditional RF limitations. This is particularly useful in urban areas with strong coverage or in applications that require real-time video streaming and cloud-based data analysis. However, cellular dependence introduces vulnerabilities such as reliance on existing infrastructure, exposure to cyber threats, and potential disruption during large-scale disasters or in contested military zones.

Another emerging area is the use of mesh networking and swarm communication protocols, where drones communicate directly with each other rather than relying solely on a central controller. These networks use short-range RF links to maintain coordinated flight, share telemetry data, and even delegate tasks autonomously. Operating on low-bandwidth but highly resilient channels, such systems are vital for drone swarms used in battlefield reconnaissance, distributed sensor networks, or coordinated attacks.

In summary, UAV RF signal control spans a wide spectrum of frequencies and technologies, from common ISM bands to highly encrypted satellite links. The choice of frequency and format depends on mission requirements, environmental conditions, and regulatory constraints. Advances in RF technologies continue to enhance UAV capabilities, pushing the boundaries of range, security, and autonomy. As UAV roles expand in both civilian and military sectors, the sophistication and diversity of RF control systems will remain central to their evolution.

The integration of Wi-Fi and 4G/5G cellular networks into unmanned aerial vehicle (UAV) control systems represents a significant evolution in drone technology, particularly in how these systems are managed, monitored, and leveraged for both civilian and military operations. Traditionally, UAVs operated using radio frequency (RF) control within line-of-sight, using point-to-point communication via handheld transmitters and receivers. While effective over short distances, this method limited UAV range and required constant human oversight. The introduction of Wi-Fi and cellular technologies has dramatically expanded the operational envelope of UAVs, allowing for beyond visual line-of-sight (BVLOS) control, real-time telemetry, high-definition video streaming, and multi-drone coordination, all with greater reliability and flexibility.

Wi-Fi is commonly used in consumer and hobbyist drones due to its widespread availability, ease of use, and ability to transmit video feeds and control signals with low latency within a range of approximately 100 to 300 meters under ideal conditions. Wi-Fi-based control is sufficient for many recreational or inspection applications, such as filming, surveying small areas, or conducting short-range reconnaissance. However, the limitations of Wi-Fi become apparent in challenging environments where interference is high, such as urban areas with dense wireless traffic, or where obstacles block direct line-of-sight between the drone and the controller. Furthermore, Wi-Fi's limited range hinders its use in long-range or mission-critical operations where consistent connectivity is paramount.

To overcome these limitations, 4G and 5G cellular networks have been adopted, especially in commercial and government UAV operations. Even in convict zones, where available, these networks offer broader coverage, lower latency, and higher data bandwidth, making them ideal for real-time video streaming, sensor data transmission, and control signal integrity over longer distances. With cellular connectivity, drones are no longer confined to the immediate vicinity of the operator, as long as they remain within the coverage area of the mobile network. This capability is especially useful in search and rescue missions, agricultural monitoring, large-scale infrastructure inspections, and military reconnaissance, where UAVs may need to operate tens of kilo-somerset further away from the base station.

6.1.3 UAV Wi-Fi and 4G/5G cellular networks

4G LTE networks provided the first significant leap in long-range UAV connectivity, enabling operators to command drones from virtually any location with mobile coverage. UAVs equipped with 4G modems can communicate with ground control stations via the internet, using standard protocols and encryption to ensure secure transmission. The reliability and ubiquity of 4G made it suitable for a wide range of applications; however, it still presented some latency issues (typically between 30–50 milliseconds) and was not optimised for the ultra-reliable low-latency communications (URLLC) that advanced drone applications require.

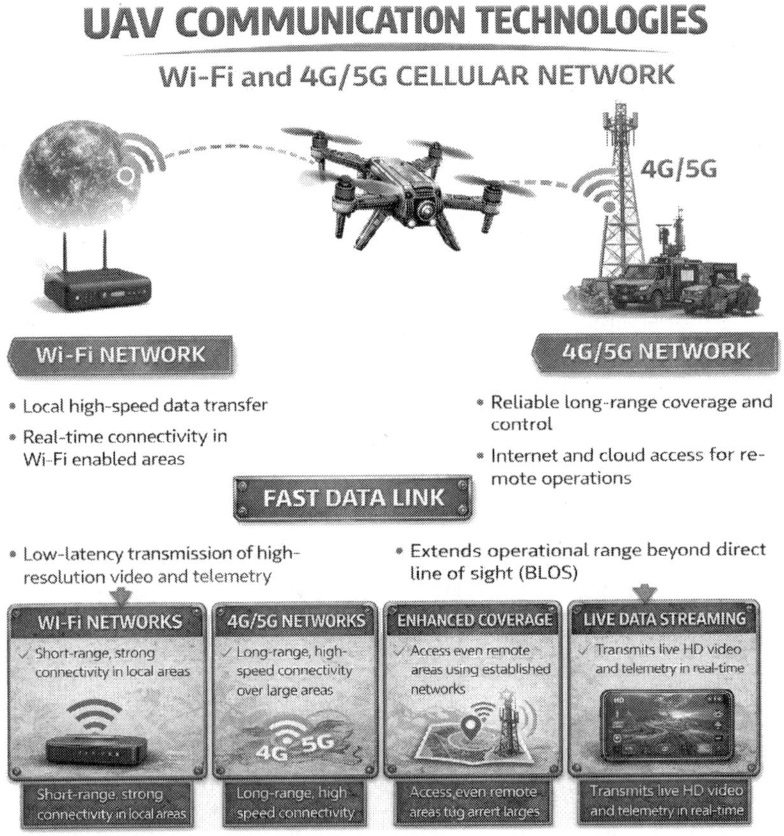

The roll-out of 5G has introduced trans-formative possibilities for UAV operations. With latency potentially reduced to below 10 milliseconds and support for enhanced mobile broadband (eMBB), 5G enables highly responsive control and simultaneous transmission of high-resolution video and sensor data. This is crucial for real-time decision-making in dynamic environments such as disaster zones, combat areas, and urban logistics. Additionally, 5G networks support massive machine-type communication (mMTC), allowing for the simultaneous connection of thousands of UAVs and ground systems, which is vital for swarm drone operations and autonomous air traffic management.

The development of UAV control systems utilising 5G has also enabled edge computing and artificial intelligence integration. Drones can now offload computationally intensive tasks such as image processing, object detection, and navigation to nearby edge servers, minimising onboard hardware requirements while maximising responsiveness and functionality. This networked intelligence paradigm significantly enhances UAV autonomy, enabling adaptive path-finding, coordinated mission execution, and AI-driven threat response in military settings.

However, the reliance on commercial cellular networks presents several challenges. First, coverage is still inconsistent in remote or undeveloped regions, which may hinder UAV deployment in rural or wilderness areas. Second, security remains a primary concern; while cellular networks use encryption and secure protocols, the risk of cyber intrusion or signal jamming is not negligible, particularly in hostile environments. In military operations, for instance, adversaries may attempt to intercept, jam, or spoof cellular communications to disrupt UAV missions. This has led to the development of hybrid communication systems, where UAVs switch between RF, Wi-Fi, and cellular networks based on signal strength, mission profile, and operational security requirements.

Furthermore, the regulatory environment must evolve to keep pace with the capabilities provided by cellular-enabled drones. Airspace authorities worldwide are working to establish standards for BVLOS operations, cellular command and control (C2) links, and network-based UAV identification and tracking. The adoption of 5G for UAVs necessitates spectrum allocation, quality of service guarantees, and infrastructure cooperation between aviation and telecommunications sectors.

In summary, the use and development of Wi-Fi and 4G/5G cellular networks in UAV control systems has revolutionised drone capabilities, enabling extended range, real-time data transmission, and integration with smart network services. While Wi-Fi remains suitable for short-range civilian applications, 4G and especially 5G cellular connectivity offer unprecedented scalability and functionality, unlocking the full potential of UAVs in complex, data-intensive, and high-risk environments. As technology continues to evolve, these networks will play a central role in the future of autonomous flight, aerial logistics, disaster response, and combat drone operations.

6.1.4 UAV Autonomous Coms (IMUs) and (GPS)

In modern warfare, the fusion of Unmanned Aerial Vehicles (UAVs) with autonomous communication systems driven by Inertial Measurement Units (IMUs) and Global Positioning Systems (GPS) has revolutionised battlefield intelligence, surveillance, and operational coordination. These two technologies form the core of UAV autonomy, allowing drones to operate with minimal human intervention while ensuring precise navigation, mission execution, and inter-vehicle communication in high-stakes, contested environments. IMUs, which include accelerometers, gyroscopes, and sometimes magnetometers, provide critical data on a UAV's orientation, velocity, and acceleration in three-dimensional space. This continuous stream of motion and positioning information enables UAVs to maintain stable flight and react immediately to dynamic changes in the environment, such as turbulence, evasive maneuvers, or GPS signal loss caused by jamming or signal denial. When paired with GPS, which delivers real-time global location data via satellite connectivity, these systems together create a powerful fusion that allows UAVs to pinpoint their position, navigate autonomously through mission waypoints, and coordinate with allied assets in complex combat zones. Sensor fusion algorithms, such as extended Kalman filters, combine IMU and GPS data to correct for drift, filter out noise, and produce accurate, reliable positioning in real time—essential for precision targeting, loitering operations, and synchronized swarm behavior in the field.

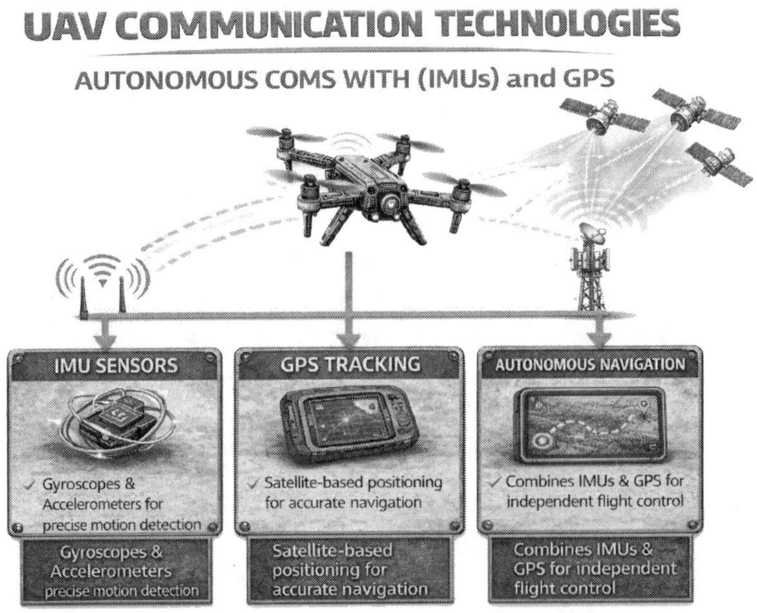

The battlefield advantage of integrating autonomous communication with GPS and IMUs lies in the UAV's ability to not only self-navigate but also share its telemetry with other units and command centers without relying on constant human direction. Through encrypted mesh networks or tactical data links, UAVs operating in swarms or collaborative reconnaissance missions can exchange IMU-derived motion data and GPS-based geolocation among themselves. This enables the formation of dynamic aerial networks that adjust formations, designate targets, or maintain area coverage with real-time awareness and minimal latency. Such autonomous coordination allows for distributed intelligence across fleets of UAVs, where each unit contributes to a shared operational picture. This is particularly valuable in denied-access areas, where centralized control may be vulnerable to disruption. GPS/IMU-fed autonomy enables fallback mechanisms where, even under electronic warfare conditions that degrade GPS reliability, UAVs can continue executing flight paths, returning to base, or executing tactical maneuvers using inertial data alone. Advanced combat UAVs may also be equipped with machine learning models trained to interpret fused GPS-IMU data and make rapid, adaptive decisions, such as re-routing based on threats or dynamically selecting new surveillance zones, enhancing mission resilience and reducing operator burden during high-tempo engagements.

Moreover, GPS and IMU synergy supports time-sensitive targeting and real-time strike coordination. In contested environments, UAVs equipped with guided munitions must rely on their onboard positioning to deliver ordnance with surgical precision. The ability to track orientation and motion—without relying solely on external signals—makes these systems difficult to spoof or disrupt, granting a decisive edge in electronic warfare scenarios. Autonomous UAVs can loiter silently above combat zones, update their location continuously, and integrate IMU data into targeting solutions even when GPS is intermittently unavailable. Additionally, UAVs conducting electronic warfare or signals intelligence missions depend on accurate geo-location to triangulate enemy positions or jam hostile communications while evading detection. IMU-guided flight paths, synced with GPS-based operational geo-fences, allow these UAVs to conduct delicate maneuvers while remaining within mission parameters. This geo-spatial precision is vital for reducing collateral damage, avoiding friendly fire, and complying with rules of engagement in complex operating theatres where applicable.

Ultimately, the marriage of GPS and IMUs in UAV platforms has ushered in a new era of autonomous capability and battlefield agility. These systems not only empower UAVs to communicate, navigate, and operate with minimal human input but also elevate the strategic potential of unmanned systems in high-threat scenarios. As advancements continue in inertial navigation accuracy, satellite resilience, and onboard AI decision-making, we can anticipate increasingly intelligent, self-reliant UAVs that function as autonomous teammates in the battle-space—filling reconnaissance gaps, extending strike reach, and enhancing command-and-control networks with unmatched precision and reliability.

6.1.5 UAV Fibre Optic Communication

Fibre optics are employed in UAV systems primarily through two key methods. The first involves tethered UAVs, which are physically connected to a ground control station using a fibre optic cable. This cable not only ensures a stable and secure communication link but can also supply continuous power, allowing for extended operation without reliance on onboard batteries. The second method applies to untethered UAVs, where fibre optics are integrated internally to facilitate high-speed data transmission between onboard components. Additionally, these drones can use retractable fibre optic systems to connect with external host platforms such as ships or vehicles, enabling reliable and interference-resistant communication without relying on wireless links.

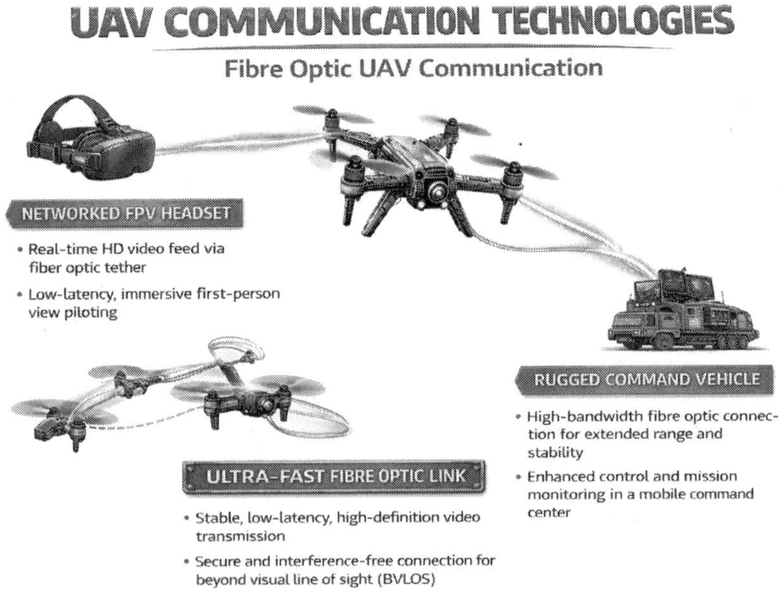

Fibre optic control has emerged as a significant evolution in UAV command architecture, particularly in contested environments where electronic warfare dominates the electromagnetic spectrum. Its immunity to electromagnetic interference (EMI) provides a decisive advantage over traditional radio frequency (RF) systems, which are vulnerable to jamming, spoofing, and interception. By transmitting data as pulses of light rather than radio waves, fibre optic lines ensure secure, uninterrupted communication between operator and platform, even in heavily saturated battle-spaces. High bandwidth capacity allows the real-time transfer of high-definition video, telemetry, and sensor data without degradation, while near-zero latency enables precise control during time-critical manoeuvres. These attributes make fibre optic systems especially valuable for missions requiring secure targeting, indoor navigation, or operations conducted under intense electronic pressure.

Fibre optic-controlled UAVs have been deployed in tethered surveillance roles, submarine-launched concepts, and ship-based observation platforms, where secure, interference-resistant communication is essential. Kevlar-reinforced tethers, multiplexed data transmission systems, and automated winch mechanisms enable stable deployment and retrieval, while immersive pilot interfaces using head-mounted displays enhance situational awareness. Hybrid concepts, including detachable fibre links and swarm networking, further demonstrate the adaptability of optical communication in modern drone operations. In these contexts, fibre optics offer speed, security, and resilience unmatched by conventional RF links.

However, the increasing use of fibre optic spools in loitering munitions and so-called "kamikaze" drones introduces a less discussed consequence: environmental pollution. Unlike tethered systems designed for recovery, many fibre-guided strike drones deploy long lengths of ultra-thin optical cable that are left behind across fields, forests, urban infrastructure, and agricultural land once the munition completes its mission. These fibre strands—often polymer-coated and reinforced with synthetic materials—do not readily biodegrade. Accumulated over repeated operations, they can create entanglement hazards for wildlife, interfere with farming equipment, and contribute to micro-plastic contamination in soil systems. In regions subjected to prolonged conflict, such as heavily contested rural zones, discarded fibre lines can become widespread, compounding the broader environmental degradation already caused by explosive remnants and industrial damage.

The pollution issue is not limited to the fibre itself. Spool casings, plastic housings, and fragmented composite components from expendable fibre-guided drones add to battlefield debris. As fibre optic-guided munitions are designed to be low-cost and disposable, recovery is rarely feasible, meaning environmental accumulation is an inherent by-product of their use. In areas where agricultural land or sensitive ecosystems are affected, the long-term ecological consequences—particularly the persistence of synthetic polymers in soil and waterways—remain insufficiently studied and will linger long into the future.

As fibre optic control continues to expand in tactical UAV applications, including strike platforms, these environmental implications require closer examination. While optical guidance offers undeniable operational advantages—resistance to jamming, secure data transmission, and precision control—the sustainability costs associated with disposable fibre deployment must be acknowledged. The future evolution of fibre-guided UAV systems may therefore need to consider alternative materials, recovery mechanisms, or improved battlefield clearance processes to mitigate the environmental footprint. In modern warfare, technological advantage and environmental responsibility are increasingly intertwined, and the proliferation of fibre optic-controlled munitions underscores the importance of addressing both dimensions simultaneously.

6.1.6 UAV Coms Systems, Long Range (SATCOM)

In long-range warfare scenarios, the effectiveness of unmanned aerial vehicle (UAV) operations heavily depends on the robustness, reliability, and security of the control and management systems used. Unlike short-range consumer drones, long-range military UAVs must operate over vast distances, often in hostile or contested environments where traditional line-of-sight (LOS) communication is impossible, and the threat of signal interference, jamming, or interception is high. The most effective control and management systems for such UAVs are those that combine satellite communication (SATCOM), advanced autonomy through onboard artificial intelligence, encrypted data links, and redundant navigation and control protocols. These integrated systems allow UAVs to perform reconnaissance, surveillance, strike missions, and electronic warfare tasks far from their operators while maintaining command integrity and mission continuity.

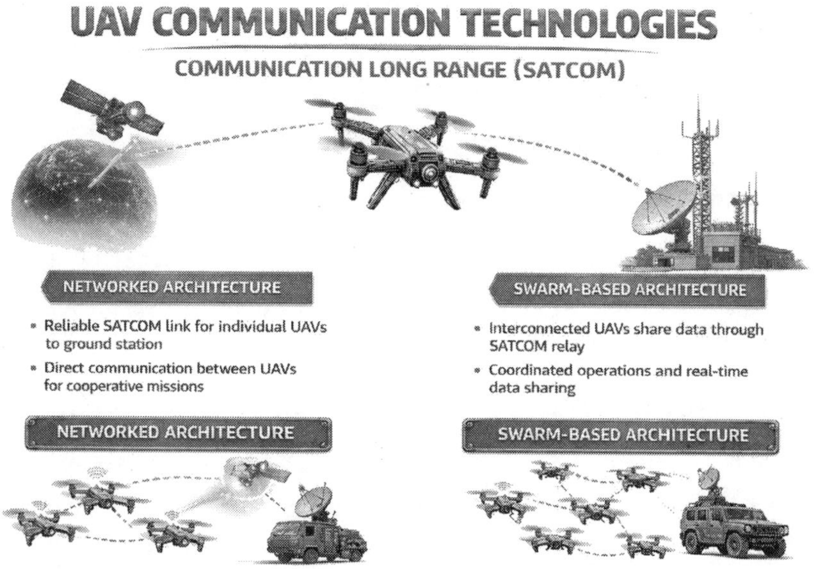

SATCOM is arguably becoming the cornerstone of long-range UAV control, in both the military and hobby drone developments. Unlike RF-based line-of-sight or Wi-Fi systems, which are limited to tens of kilometres at best, SATCOM provides global coverage, enabling UAVs to operate thousands of kilometres from their launch or control stations. Military UAVs such as the MQ-9 Reaper or RQ-4 Global Hawk use satellite links to transmit high-resolution video feeds, sensor data, and telemetry to ground control stations in near-real time. These links are essential for precision targeting, battlefield monitoring, and post-strike assessment. SATCOM not only supports direct pilot control but also facilitates command-and-control (C2) relay, allowing operators to redirect missions mid-flight or pass control between ground stations as the UAV transits across regions. However, satellite systems are expensive and require infrastructure such as ground stations, satellite time, and encryption protocols, which are typically available only to state actors or well-funded paramilitary groups.

To bolster the effectiveness of SATCOM-based systems, modern UAV control architectures often incorporate autonomous mission management and onboard artificial intelligence. These autonomous capabilities are essential when operating in GPS-denied or communication-disrupted environments. AI-enabled systems can interpret sensor inputs in real time, make flight adjustments, reroute around threats, and complete tasks without waiting for operator commands. For instance, a UAV tasked with monitoring enemy troop movements may autonomously shift its path if it detects radar activity or anti-aircraft missile launches. Autonomy also reduces operator workload and latency in decision-making, which is critical during time-sensitive missions like high-value target elimination or dynamic battlefield reconnaissance.

Another vital aspect of effective long-range UAV control is the use of secure, redundant, and encrypted communication links. In warfare, adversaries frequently attempt to jam or spoof UAV signals to either disable them or hijack control. To counteract this, military-grade UAVs use frequency-hopping spread spectrum (FHSS) and encrypted communication protocols, often layered with multiple backup systems. For example, a UAV might simultaneously transmit telemetry via SATCOM and low-bandwidth high-frequency (HF) radio as a backup. If the primary satellite link is compromised, the UAV can downshift to an alternate communication mode or even enter autonomous "return to base" or "mission continuation" protocols based on its programming. This redundancy is crucial for operational survivability and mission assurance.

Control and management systems are also increasingly adopting networked or swarm-based architecture, where multiple UAVs communicate with each other and share data dynamically. In this distributed model, one UAV losing communication does not jeopardise the mission, as another can assume command or relay information. Such mesh networks, combined with edge computing, allow for resilient, decentralised control—particularly effective in environments with contested electronic domains. For example, in a long-range strike mission involving multiple UAVs, the fleet can coordinate via encrypted short-range links, adjust roles autonomously, and maintain situational awareness without constant ground oversight. This approach not only increases efficiency but also makes the drone fleet more difficult to disrupt.

Energy management and command sustainability are also key components of long-range UAV control systems. Since these drones often fly for 24 hours or more, real-time power consumption monitoring, thermal management, and fail-safe system diagnostics must be integrated into the control software. UAVs must be able to monitor their systems, detect engine or battery anomalies, and adjust mission parameters accordingly to avoid failure far from base. Ground control stations thus serve not only as pilot terminals but also as mission operation centers, receiving a stream of performance data and using predictive algorithms to make recommendations or initiate recovery plans.

In summary, the most effective control and management systems for long-range warfare UAVs are those that combine global SATCOM coverage with onboard autonomy, secure and redundant communication links, and integrated fleet or swarm networking capabilities. These systems allow drones to operate deep in enemy territory, persist for extended durations, and full-fill complex missions with minimal human input. As warfare evolves into increasingly asymmetrical and electronically contested domains, the ability of UAVs to maintain command integrity, adapt to changing circumstances, and remain operational despite communication losses or attacks will define their strategic value. Future developments will likely focus on enhancing AI autonomy, quantum encryption for globally available data links, and collaborative intelligence between manned and unmanned assets, ensuring UAVs remain a decisive factor in modern and future warfare.

6.1.7 UAV Mesh Networking, Swarm Coms

UAV mesh networking and swarm communication protocols represent a critical advancement in the evolution of unmanned aerial systems (UAS), enabling greater resilience, autonomy, and coordination in complex environments. These technologies allow multiple drones to communicate directly with each other rather than relying solely on centralised control from a ground station or satellite link. This decentralised approach significantly enhances the effectiveness and survivability of UAV operations, particularly in contested or denied environments where traditional command and control structures may be disrupted or impractical. Through mesh networking and swarm protocols, drones can form intelligent, adaptive collectives capable of executing coordinated missions such as surveillance, search and rescue, electronic warfare, and kinetic strikes with remarkable efficiency and flexibility.

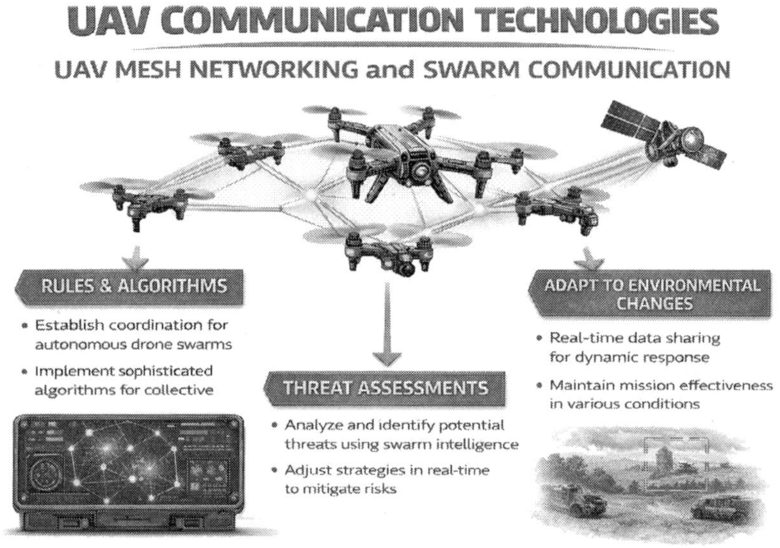

At the core of UAV mesh networking is the concept of peer-to-peer communication. Each UAV in the network acts as both a node and a relay point, allowing data to be passed dynamically between aircraft. If one drone is out of direct range of the ground control station, it can route its data through neighbouring drones to maintain connectivity. This self-healing capability means that even if some drones are lost or jammed, the network can reconfigure itself to maintain communication and mission continuity. Mesh networking protocols also enable real-time sharing of telemetry, sensor data, position, mission updates, and command instructions, creating a unified situational awareness across the swarm.

Swarm communication protocols are specialised sets of rules and algorithms that govern how UAVs interact within a mesh network to achieve collective behaviour. Inspired by biological swarms—like those of bees, birds, and fish—these protocols allow UAVs to operate cooperatively without direct human intervention.

Swarm logic dictates how drones maintain formation, avoid collisions, divide tasks, and adapt to environmental changes or threats. For instance, in a surveillance mission, swarm communication might assign drones to sweep different sectors of a battlefield, share findings instantly, and reposition in real time based on detected anomalies or priorities. In a combat setting, such as loitering munition operations, swarms can disperse upon detecting a threat, reassemble for an attack, and even assign a drone to sacrifice itself if necessary, all without human direction.

There are several key elements to effective swarm communication protocols. First is localisation and positioning. UAVs need to know their own location and that of their peers to coordinate movements and avoid interference. This is typically managed through GPS in open environments, but in GPS-denied zones, drones may use inertial navigation, vision-based localisation, or relative positioning through signal triangulation. Second is consensus and task allocation. Swarm algorithms such as consensus-based bundle algorithms (CBBA), market-based models, and distributed task assignment allow drones to negotiate roles dynamically. For example, if one drone detects a high-value target, the swarm can assign the most suitable unit for close inspection or engagement, based on proximity, fuel levels, and payload status.

Communication within a UAV swarm relies on robust, low-latency, and often encrypted wireless links. Common technologies include Wi-Fi, LoRa, and increasingly, military-grade radios with frequency-hopping spread spectrum (FHSS) and ultra-wideband (UWB) capabilities to resist jamming and interception. For tactical applications, mesh networks operate in dynamic spectrum environments, scanning available frequencies and switching automatically to avoid interference. Protocols like Mobile Ad-hoc Network (MANET) standards and Optimised Link State Routing (OLSR) or Ad-hoc On-demand Distance Vector (AODV) routing protocols are commonly used to ensure data packets find the most efficient path through the network.

One of the most trans-formative developments in this field is the integration of artificial intelligence and machine learning into swarm protocols. AI allows the swarm to adapt its behaviour based on mission objectives, environmental feedback, and adversarial actions. For example, if a portion of the swarm is destroyed or jammed, the remaining drones can reassess the situation and reform the mission plan autonomously. In reconnaissance missions, AI can be used to identify areas of interest and assign drones to perform focused scans, while others maintain over-watch or communication relay functions. This level of decentralised intelligence reduces the need for constant operator input and allows for operations in environments where communication with a central controller is unreliable or impossible.

Despite its advantages, UAV mesh networking and swarm communication face several challenges. Bandwidth limitations, signal latency, interference, and power consumption are all significant constraints, especially as swarm sizes grow into dozens or hundreds of units. Managing data flow, ensuring synchronisation, and maintaining secure links across a distributed, mobile network is technically demanding. Additionally, swarm behaviour must be predictable and safe, particularly when operating in civilian airspace or urban environments where the potential for collateral damage or interference with manned aircraft is high. Regulatory frameworks are still evolving to account for autonomous swarm operations, and international laws may need to be updated to reflect the implications of AI-driven combat drones and coordinated UAV actions.

In conclusion, UAV mesh networking and swarm communication protocols are redefining what is possible in aerial robotics and autonomous warfare. These technologies allow for decentralised, intelligent, and resilient operations that can function in environments previously inaccessible to traditional UAV systems. By enabling real-time collaboration, autonomous task execution, and adaptive mission planning, drone swarms promise to become a cornerstone of future military technology, representing the most significant threat in warfare when combined with other platforms totally overwhelming their enemy. As AI, communication hardware, and swarm algorithms continue to advance, the effectiveness, versatility, and operational reach of UAV swarms will only grow, reshaping the battlefield and beyond.

CHAPTER 10

UAV Piloting and Flight Control

7.1 Piloting and Flight Control

UAV piloting and control types refer to the various ways unmanned aerial vehicles (UAVs) are operated, ranging from direct manual control to fully autonomous operations. These methods vary in complexity, technological requirements, and the level of human intervention. Below are the primary UAV piloting and flight control methods used across civilian, commercial, and military contexts.

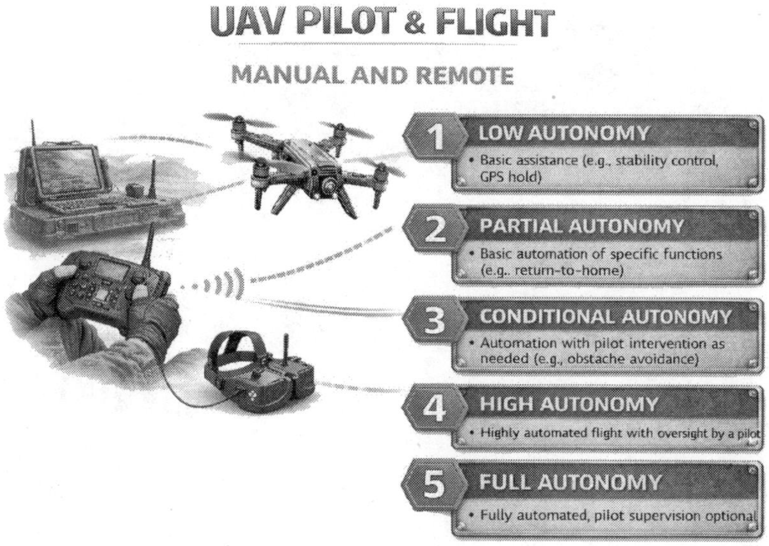

7.1.1 Manual (Remote) Control

Manual (remote) control is the most fundamental and widely recognised method of UAV operation, involving direct real-time control by a human pilot. In this mode, the UAV is guided using a handheld radio transmitter or, in more advanced settings, a ground control station (GCS). The control is typically conducted within visual line-of-sight (LoS), where the operator uses joystick inputs to navigate the UAV, much like flying a traditional remote-controlled model aircraft. This method remains especially common in hobbyist environments, where recreational users operate drones for leisure, photography, or racing. It is also prevalent in first-person view (FPV) drone racing, where precision, quick reflexes, and direct control are essential. Furthermore, manual control is sometimes employed in close-range tactical scenarios where immediate responsiveness and pilot discretion are required.

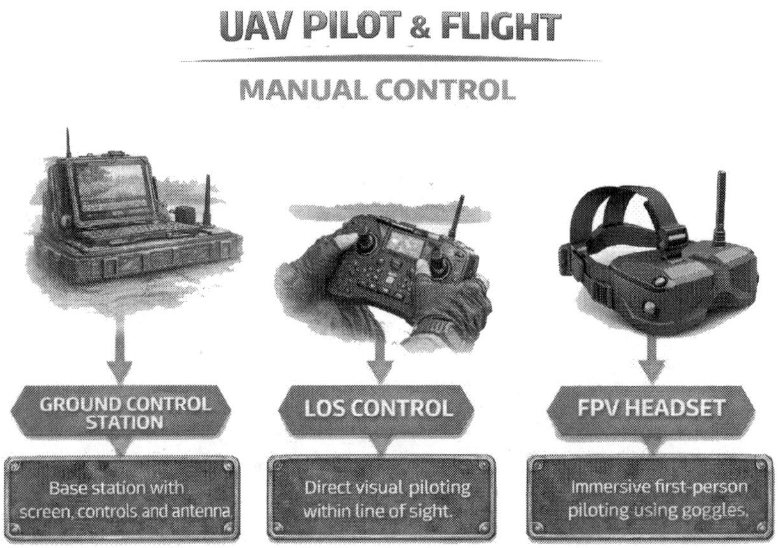

One of the main advantages of manual control is the direct and instantaneous response it offers. The pilot can immediately react to changes in the environment, such as unexpected obstacles, shifting weather conditions, or evolving mission parameters. This makes manual operation highly suitable for training purposes, enabling new pilots to develop essential flying skills and judgement. The simplicity of the setup also appeals to users in non-professional settings, as it typically requires minimal technological infrastructure compared to more automated systems.

However, manual control also presents several limitations. The need for constant human attention can lead to fatigue during prolonged operations, increasing the risk of operator error. The range is constrained by the line-of-sight requirement and the limitations of radio frequency transmission, often restricting use to a few kilometres or less, depending on terrain, antenna strength, and environmental interference. Moreover, manual piloting lacks the precision and repeatability of autonomous systems, which can be crucial in applications such as surveying, mapping, or automated delivery. Human error remains a significant concern, as lapses in judgement or delayed reactions can result in crashes, loss of equipment, or unintended intrusions into restricted airspace.

In summary, while manual (remote) control offers immediacy, flexibility, and accessibility—especially for entry-level users and in dynamic scenarios—it is inherently limited by human performance and communication range. As a result, it is best suited for short-range, real-time applications where simplicity and responsiveness outweigh the need for extended operational range or autonomous functionality.

7.1.2 Assisted (Stabilised) Manual Control

Assisted manual control is a widely used UAV piloting method that blends human input with onboard technological aids to enhance flight stability and safety. In this mode, the UAV is still operated manually by a human pilot, but advanced sensors and software features assist in maintaining altitude, orientation, and overall flight balance. These built-in systems reduce the workload on the pilot by automatically correcting for wind drift, sudden movements, or unintentional control errors, allowing for smoother and more controlled flight. Common examples of such assistance include GPS hold, which enables the drone to maintain a fixed position using satellite navigation; altitude hold, which keeps the UAV flying at a consistent height regardless of minor throttle changes; and headless mode, which simplifies directional controls by orienting the UAV relative to the pilot rather than the drone's own front.

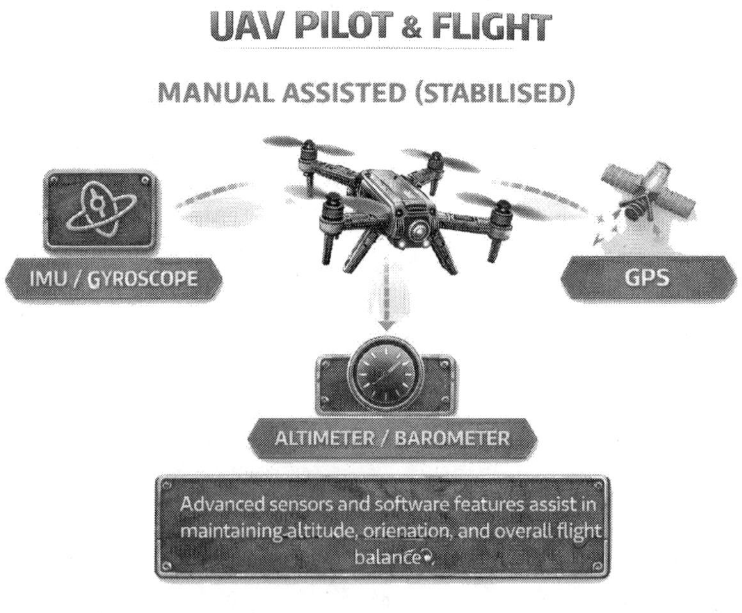

This form of control is especially popular in consumer drones and is commonly found in models used for aerial photography, video-graphy, and general surveillance. By automating basic flight stabilisation, assisted manual control makes drones significantly easier to fly, especially for beginners with limited piloting experience. This reduces the likelihood of crashes or erratic flight behaviour, which is particularly beneficial when operating expensive equipment or conducting flights in confined or obstacle-rich environments. The enhanced stability is also invaluable for tasks that demand smooth and steady footage, such as cinematic filming or high-resolution aerial mapping, where even slight movements can impact data quality or video clarity.

Despite its many benefits, assisted manual control still relies heavily on human decision-making for navigation and mission execution. The pilot is responsible for setting the course, avoiding obstacles, and responding to unexpected environmental changes, such as wind gusts or signal interference. As a result, while these assistive technologies can greatly reduce pilot error and enhance safety, they do not eliminate the need for attentiveness and a foundational understanding of UAV operation. In situations requiring complex maneuvers or rapid adjustments, the limitations of assisted control may become apparent, particularly if the pilot lacks experience or if the software fails to respond appropriately to unusual conditions.

Overall, assisted manual control offers a balanced approach between manual piloting and automation. It provides a valuable safety net and smoother flight experience without fully removing the pilot from the loop, making it ideal for a wide range of applications—from recreational use to professional drone-based services—where precision and user-friendliness are paramount.

7.1.3.Semi-Autonomous Control

Semi-autonomous control represents a hybrid approach to UAV piloting, where human operators define specific parameters—such as waypoints, altitude, speed, and mission duration—and the UAV then follows these instructions automatically. Unlike fully manual or fully autonomous systems, semi-autonomous control allows for human oversight and intervention at any stage of the operation. The pilot can take control instantly if conditions change or the drone behaves unexpectedly, making this method both flexible and reliable. Semi-autonomous control is typically implemented through user-friendly applications or Ground Control Station (GCS) software platforms such as Mission Planner, DJI Ground Station, or QGroundControl. These interfaces allow users to design flight paths visually on a map, input altitude settings, and configure mission objectives with precision, which the UAV then executes with minimal real-time input from the operator.

This control method is widely used in sectors that require routine, repeatable, and data-intensive operations. In agriculture, for example, semi-autonomous drones can be programmed to fly over large fields to conduct crop health analysis, spray fertilisers or pesticides, and collect multi-spectral imagery. Industrial inspection tasks, such as monitoring pipelines, power lines, or wind turbines, also benefit from semi-autonomous control because it allows UAVs to follow exact routes repeatedly, ensuring that inspections are consistent over time. Tactical surveillance in military and security operations also leverages this system, where drones can patrol predefined perimeters or fly reconnaissance missions autonomously while allowing human operators to intervene if a threat or anomaly is detected.

The advantages of semi-autonomous control are significant. By automating the majority of the flight process, this method dramatically reduces the pilot's workload, enabling greater focus on data collection, analysis, or mission supervision. It enhances accuracy and consistency, especially in repeated missions where precise flight paths and consistent conditions are critical. Additionally, it improves safety, as built-in safeguards like obstacle avoidance and return-to-home functions can be programmed into the mission parameters.

However, despite its many strengths, semi-autonomous control still demands human oversight. The operator must remain available to monitor the mission, respond to unexpected obstacles, communication issues, or changing weather conditions, and make quick decisions in the event of system failure. Effective contingency planning is essential, as over-reliance on automation without adequate preparation for manual takeover can lead to mission failure or loss of the UAV. In conclusion, semi-autonomous control offers a balanced and efficient solution for complex UAV operations, combining the benefits of automation with the safety and flexibility of human intervention.

7.1.4 Fully Autonomous Control

Fully autonomous control represents the most advanced level of UAV operation, where the drone is capable of completing entire missions without any real-time input from a human operator. This includes all phases of flight—takeoff, navigation, data acquisition, mission execution, and landing—conducted entirely through pre-programmed instructions or dynamic, AI-guided decision-making. These UAVs rely heavily on a suite of integrated systems, including Global Positioning System (GPS) modules for geolocation, inertial navigation systems (INS) for stability and orientation, onboard sensors for obstacle detection and avoidance, and sophisticated artificial intelligence algorithms that enable real-time adaptations to changing conditions. With these capabilities, the drone can assess its environment, avoid obstacles, and make independent flight decisions based on pre-set mission parameters or situational awareness.

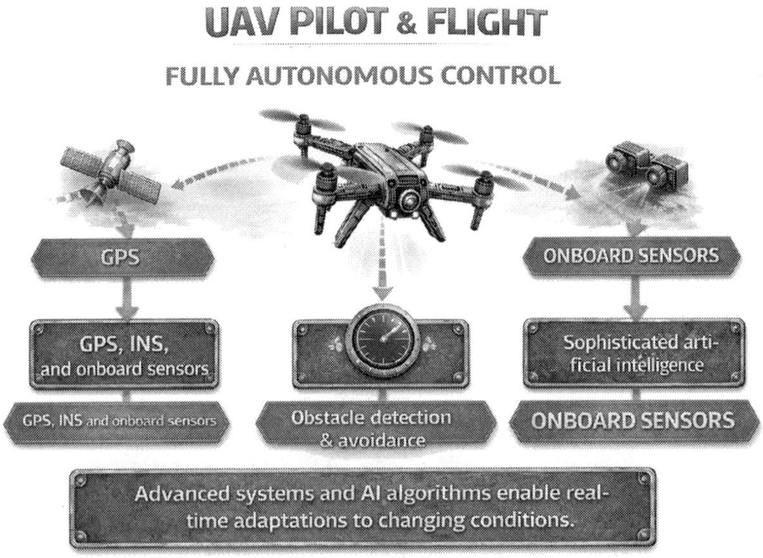

Fully autonomous UAVs are increasingly deployed in environments where direct human control is either impractical or exposes operators to unacceptable risk. In military operations, they conduct reconnaissance and long-range surveillance missions deep within hostile or denied territory, reducing the need for personnel to enter contested zones. As autonomy has advanced, these systems have also been adapted for high-risk strike roles, including long-range loitering platforms capable of remaining over target areas for extended periods before engagement. In such scenarios, accessibility may be limited, infrastructure damaged, or electronic interference pervasive. Autonomous systems provide consistent monitoring of terrain, infrastructure, and troop movement, performing repeatable, large-scale tasks with precision and endurance beyond human capability.

The operational advantages of full autonomy are considerable. Autonomous UAVs can function continuously, unaffected by fatigue, stress, or degraded situational awareness. They can follow pre-programmed or dynamically adjusted flight paths with high precision, maintaining consistent mission parameters across repeated deployments. Their scalability further enhances force multiplication; coordinated fleets can operate simultaneously across wide operational areas, managed through centralised command architectures with limited human intervention. This capability is particularly valuable in data-intensive environments where persistent coverage and pattern recognition are essential.

However, the transition to fully autonomous control introduces significant technical and operational challenges. These systems remain vulnerable to external disruption, including GPS spoofing, signal jamming, or cyber intrusion, all of which can compromise navigation or mission integrity. The absence of real-time human oversight requires highly reliable onboard processing, redundant navigation systems, and advanced fail-safe mechanisms to mitigate unintended consequences. Sensor malfunction, corrupted data inputs, or flawed AI-driven assessments can result in mis-identification, navigation errors, or uncontrolled flight behaviour. Consequently, the deployment of fully autonomous UAVs demands rigorous validation, layered redundancy, and comprehensive safeguards to ensure operational reliability in unpredictable and electronically contested environments.

At the same time, rapid advances in open-source artificial intelligence and machine learning (ML) have accelerated the evolution of autonomous strike capabilities. Modern systems are increasingly trained on extensive datasets to recognise patterns, classify objects, and distinguish hostile assets from background activity. In targeted applications, ML-driven models analyse sensor inputs in real time, identifying potential threats based on behavioural signatures, movement patterns, and contextual indicators. Once validated against programmed engagement criteria, onboard AI systems can execute terminal guidance sequences within seconds. This fusion of large-scale data modelling with automated decision architecture has enabled autonomous strike platforms to operate in heavily contested environments where traditional, manually controlled systems would struggle to function. As autonomy continues to mature, the integration of AI and ML into targeted engagement systems is reshaping the tempo and character of modern conflict.

7.1.5 (FPV) and Tethered Control

First-Person View (FPV) control is a UAV piloting method where the operator navigates the drone using a live video feed transmitted from an onboard camera, viewed through specialised FPV goggles or a ground-based monitor. This immersive control style offers a pilot's-eye perspective, allowing the operator to see exactly what the drone sees in real time, which enhances spatial awareness and responsiveness. FPV control transforms the flight experience from external observation to internal navigation, enabling highly precise maneuvering in environments that may be too complex or constrained for traditional line-of-sight control. It is especially popular in drone racing, where speed, agility, and instant reaction times are critical. The system allows pilots to perform high-speed aerial maneuvers with a level of precision that would be difficult or impossible from an external viewpoint.

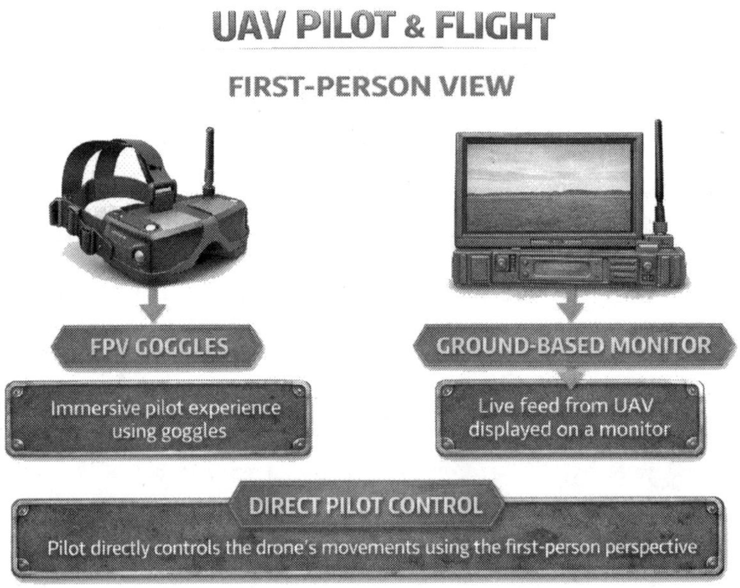

The primary advantage of FPV (first-person view) control lies in the high level of situational awareness it provides to the operator. By delivering a live visual feed directly from the drone's perspective, FPV enables precise navigation and informed decision-making in fast-moving, cluttered, or visually complex environments. This immersive control method allows pilots to thread through obstacles, visually confirm and track hostile targets, and maintain accurate alignment during terminal or "hard-kill" phases of an engagement. Reaction times are significantly improved, as operators perceive threats and required course corrections instantly, without the spatial and cognitive delay associated with ground-referenced piloting.

Despite these strengths, FPV systems have traditionally been constrained by limited operational range. Because FPV relies on continuous, low-latency video transmission, maintaining a stable signal is essential;

any disruption, lag, or interference can rapidly degrade control authority. Urban terrain, vegetation, structures, and elevation changes frequently obstruct signal paths, while analogue and digital RF-based FPV links remain vulnerable to noise, jamming, and limited penetration in dense or indoor environments. Although, longer fibre optic tethers, infrared links, and advanced relay architectures can extend range and improve signal fidelity in specialised roles, they introduce added system complexity and can restrict manoeuvrability or deployment flexibility.

Many of the historical limitations associated with FPV goggles and short-range links are now being mitigated through the use of extended fibre optic connections and high-end satellite communications (SATCOM) systems. These technologies enable operators to retain immersive, real-time control over significantly greater distances, even in electronically contested battle-spaces. However, such solutions introduce their own constraints, including logistical burden, susceptibility to physical damage, and environmental pollution concerns—particularly where disposable fibre links are employed in strike applications. Nevertheless, as the requirement to overcome hostile targets in denied environments becomes increasingly paramount, the integration of long-range FPV interfaces with resilient communications will continue to expand, reinforcing FPV control as a critical element in precision engagement and modern unmanned combat operations.

Tethered UAVs

Similar to FPV and fibre optic control systems, tethered UAV control can be considered within the same broader category of direct, real-time operator-managed flight architectures.

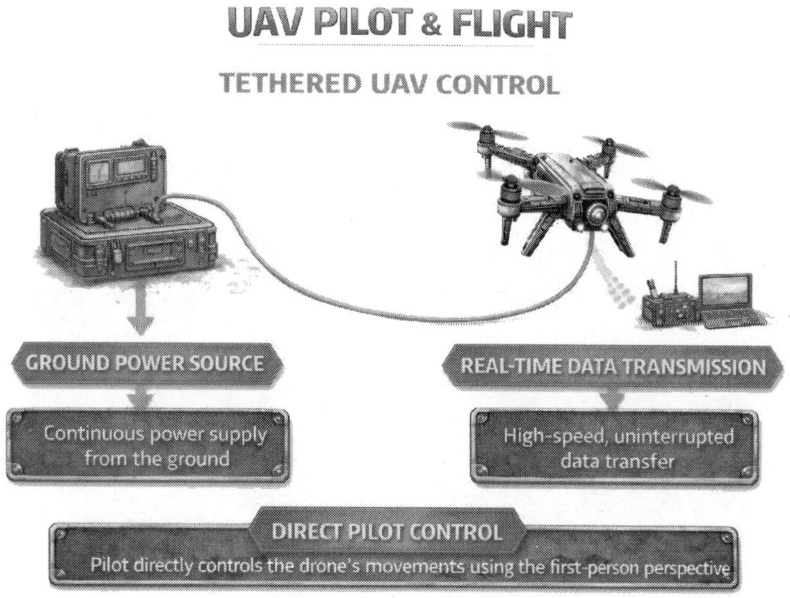

Similar to FPV and fibre optic-controlled systems, tethered UAV control can be placed within the same general category of direct, real-time operator-managed flight architectures. In many configurations, a tethered UAV may also incorporate a hardened fibre optic lines within the cable itself, enabling secure data transmission alongside physical linkage. However, this approach is typically reserved for specific operational scenarios where both the aircraft and tether can be safely recovered, as the drone remains physically connected to a ground station or vehicle-mounted control unit. Because the cable limits manoeuvre radius, tethered systems are generally deployed over shorter distances and in controlled environments rather than for deep penetration missions.

Tethered UAVs are most commonly used for surveillance and reconnaissance roles where persistent aerial presence is more important than mobility. The tether supplies a continuous flow of electrical power from a ground-based source, removing reliance on onboard batteries or fuel reserves. This effectively eliminates one of the primary constraints of conventional free-flying drones—limited endurance—allowing tethered platforms to remain airborne for extended or even theoretically unlimited durations. Such sustained flight capability makes them particularly valuable for perimeter monitoring, border security, convoy over-watch, and fixed-site protection.

Another significant advantage lies in the strength and stability of the data connection. Unlike wireless RF links, which can be disrupted by interference, jamming, or signal attenuation, a wired tether provides a secure and consistent communication pathway. This ensures uninterrupted transmission of high-resolution video, telemetry, and control data, which is especially critical in tactical military environments where real-time situational awareness and data integrity are paramount. The combination of sustained power delivery and robust data security positions tethered UAV systems as a reliable solution for missions requiring endurance, stability, and resistance to electronic disruption.

7.1.6 Beyond Line-of-Sight (BLoS) Control

Beyond Line-of-Sight (BLoS) control is a sophisticated method of UAV operation that enables the pilot or mission controller to manage the drone from long distances—often well outside the visual or radio range—by utilising advanced communication technologies such as satellite links, cellular networks (4G/5G), or airborne relay systems. This capability removes the spatial limitations of traditional line-of-sight control, allowing UAVs to operate across regional, national, or even global distances without the need for direct visual or radio contact. BLoS control has become increasingly important in modern UAV applications, particularly in military operations, where long-endurance drones are used for reconnaissance, surveillance, and precision strikes in foreign or hostile environments. It is also widely applied in large-scale commercial and logistical uses, such as drone delivery systems that cover vast rural or urban territories, and in national security operations like border patrol, where drones must cover long stretches of remote terrain.

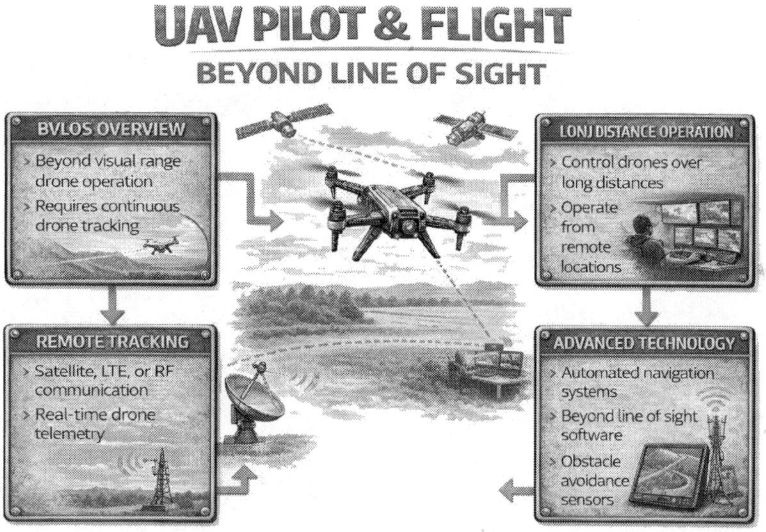

The primary advantage of BLOS, (Beyond line of sight) control is that it significantly extends the operational reach and mission versatility of UAVs. Drones can be deployed from a central hub and managed remotely from thousands of kilometres away, enabling real-time mission adjustments and live telemetry updates without requiring a local ground crew. This is especially useful for endurance platforms like the MQ-9 Reaper or maritime surveillance UAVs, which can remain airborne for more than 24 hours while transmitting intelligence data to operators in secure, distant control centres. In commercial settings, BLoS capability allows for the expansion of automated drone services, including emergency medical supply delivery to remote areas and wide-area environmental monitoring.

However, operating drones BLoS introduces several technical and operational challenges. One of the most significant limitations is the reliance on complex infrastructure. SATCOM systems require satellite availability, dish alignment, and robust ground stations, while cellular-based BLoS requires uninterrupted coverage from commercial towers, which may be unavailable in remote or mountainous areas. Moreover, these communication systems are vulnerable to disruption. GPS spoofing, jamming, or cyber attacks can interfere with data transmission or cause loss of control. Basic radio or non-hardened links may not be sufficient for critical missions, necessitating the use of more sophisticated and encrypted communication protocols. In environments where reliability and security are paramount, particularly in military or sensitive commercial operations, UAVs operating BLoS must be equipped with high-integrity, fail-safe communications systems and autonomous fallback procedures to maintain safety and mission continuity. As a result, while BLoS dramatically expands UAV capability, it also demands a higher level of technical maturity and investment.

7.1.7 Swarm Control

Swarm control represents a trans-formative leap in the deployment and coordination of unmanned aerial vehicles (UAVs), where a single operator or autonomous algorithm manages a networked group of UAVs functioning collectively as a cohesive unit. Drawing inspiration from natural swarm behaviour—like that of flocks of birds or schools of fish—this approach leverages advanced artificial intelligence (AI), decentralised decision-making, and mesh communication networks to enable real-time synchronisation among multiple drones. Rather than acting as isolated agents, each UAV in the swarm operates as part of a larger collective system, responding to dynamic changes in the environment or mission objectives in a distributed yet coordinated fashion.

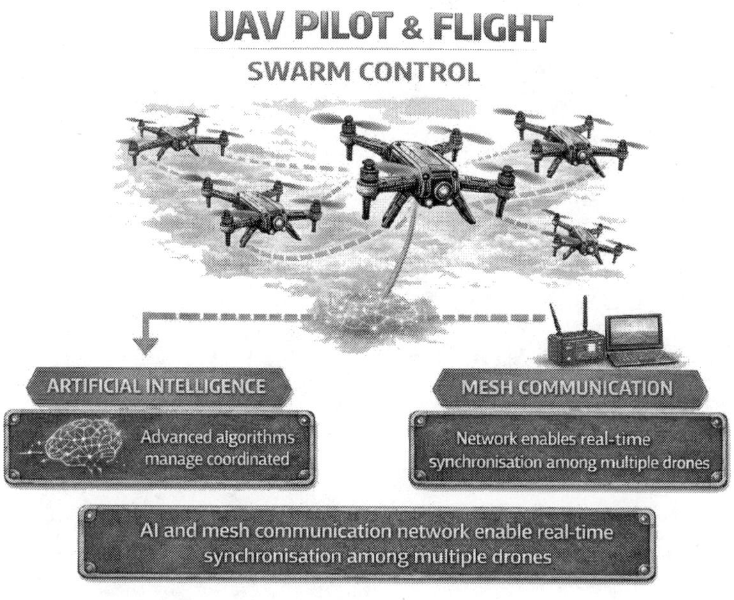

This operational model is increasingly being utilised across a spectrum of complex applications, notably in military strike scenarios where simultaneous targeting from multiple vectors can overwhelm enemy defences. Similarly, in reconnaissance and combined missions, drone swarms offer broad coverage across difficult terrain, rapidly scanning vast areas to locate targets or assess hazardous zones. Monitoring the enemy is another key field of application, where coordinated UAVs can simultaneously gather data on combat conditions, enemy positions, or troop movements across large and otherwise inaccessible regions.

One of the standout advantages of swarm control is its scalability. As mission demands grow, additional UAVs can be seamlessly added to the network with minimal disruption, allowing for flexible expansion of operational capacity. The collective intelligence of the swarm provides redundancy as well; if a single unit fails or is compromised, the remaining drones can dynamically compensate for its loss, ensuring continued mission effectiveness. This redundancy boosts resilience and reduces reliance on any single point of failure. Furthermore, swarm intelligence enhances situational awareness and responsiveness, as drones share sensor data and compute collaboratively to make group decisions on-the-fly.

However, realising the full potential of UAV swarming comes with its share of challenges. "Sophisticated AI algorithms are required" to govern swarm behaviour, and robust mesh networking infrastructure must be in place to facilitate seamless communication between units. The complexity of coordinating multiple autonomous agents in real time raises concerns about communication latency and signal interference, which could hinder effectiveness in high-stakes environments. Moreover, ensuring security and preventing adversarial exploitation of networked systems is a critical consideration.

Despite these hurdles, the capabilities introduced by swarm control have opened up bold new possibilities in UAV operations. As AI and network technologies continue to evolve, we can expect increasingly intelligent, adaptive, and reliable UAV swarms to play a prominent role in shaping the future of aerial warfare, engagements and drone developments.

7.1.8 Manned-Unmanned Teaming (MUM-T)

Manned-Unmanned Teaming (MUM-T) is an evolving concept in modern warfare that integrates the strategic advantages of human operators with the extended capabilities of unmanned aerial vehicles (UAVs). In this system, piloted aircraft or ground-based controllers work collaboratively with one or more UAVs, directing their missions and receiving real-time data through sensor feeds. This collaborative structure allows for an extension of the pilot's situational awareness and decision-making range, transforming how military operations are conducted in the battle-space. MUM-T empowers human operators to leverage the mobility, persistence, and sensor reach of unmanned systems while maintaining command over complex, high-stakes scenarios. In tactical environments, especially during wartime engagements, this fusion of human oversight with machine autonomy enhances mission success by enabling coordinated strikes, expanded reconnaissance coverage, and greater flexibility in executing operations across diverse and rapidly shifting terrains.

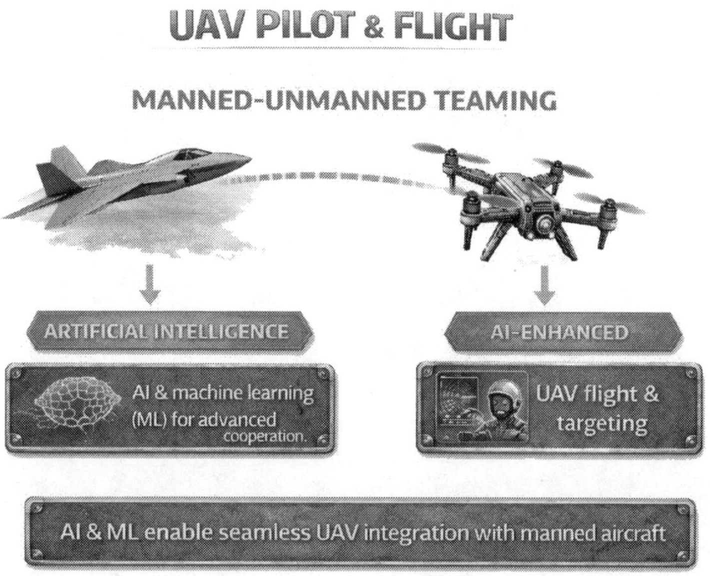

A central benefit of MUM-T in combat missions is the significant enhancement of operational reach without placing human lives directly in harm's way. UAVs can be deployed into high-risk zones to perform reconnaissance, electronic warfare tasks, or targeted engagements, all while transmitting live intelligence back to the manned platform. This enables pilots to make better-informed decisions with a broader field of vision and an additional layer of tactical advantage. The manned platform essentially becomes a battlefield command hub—situated within or near the combat environment—capable of orchestrating multiple remote assets that act as its eyes, ears, and extended reach. In scenarios where time and accuracy are paramount, such as deep strike missions or close air support, the synergy provided by MUM-T helps compress the kill chain and increase responsiveness to threats or emerging targets.

However, the sophistication that makes MUM-T effective also introduces serious challenges. Effective MUM-T operations rely on robust interoperability between manned and unmanned systems, requiring seamless integration of communication protocols, data formats, and control interfaces. In combat zones, where electronic countermeasures, cyber threats, and signal interference are real concerns, uninterruptible communications become not just advantageous but critical. Furthermore, as the number of UAVs in a single mission grows, managing and interpreting multiple streams of sensor data requires advanced artificial intelligence and machine learning algorithms. These technologies must be capable of autonomous decision support, task allocation, and adaptive behaviour under fluid combat conditions, reducing operator workload and ensuring mission continuity even when human attention is divided. While complex, the potential of MUM-T to revolutionise military effectiveness is undeniable, and ongoing advancements in AI, secure communications, and systems interoperability continue to push the boundaries of what this powerful team dynamic can achieve on the modern battlefield.

7.1.9 Summary, Piloting & Flight Control

In summary, Unmanned Aerial Vehicle (UAV) piloting and flight control systems encompass a wide array of technologies and methodologies, each tailored to meet the evolving demands of different missions and operational environments. From the simplicity of direct line-of-sight manual control, where pilots navigate UAVs with radio transmitters much like model aircraft, to the complexity of autonomous, AI-driven swarm operations capable of orchestrating dozens—or even hundreds—of drones as a single distributed intelligence, the range of options is vast and continually expanding. In between these extremes lie a rich assortment of piloting modalities including way-point navigation, teleportation via satellite links, GPS-guided automated flight, and semi-autonomous mission execution with human supervision. These varying control types are not one-size-fits-all; each comes with distinct strengths and trade-offs that must be weighed against operational needs and mission parameters. Simpler manual systems may be ideal for short-range, low-risk surveillance tasks in urban areas where real-time visual oversight is essential, whereas more sophisticated systems leveraging AI, computer vision, and real-time data fusion may be essential for large-scale environmental monitoring, border patrol, or battlefield operations in contested airspace. The selection of the appropriate control method also hinges heavily on the availability and reliability of communication infrastructure. For instance, remote missions in GPS-denied environments or electromagnetic-contested zones may necessitate more autonomous capabilities to mitigate potential disruptions to connectivity. Operational risk is another major determining factor; high-risk scenarios—whether due to environmental hazards, hostile territory, or the critical nature of the mission—often favour systems that reduce pilot workload, offer greater redundancy, or allow for stand-off engagement via UAVs, minimising human exposure. Furthermore, the nature of the payload, whether it's high-resolution imaging sensors, communications relay equipment, or tactical munitions, influences control requirements and decision cycles. As UAV technology continues to evolve, integrating enhanced onboard processing power, secure communication protocols, and AI-driven autonomy, flight control systems are increasingly converging on hybrid models—blending human oversight with machine-driven responsiveness to create a flexible, resilient command structure. In conclusion, effective UAV piloting and flight control demand a holistic approach that considers the full spectrum of mission demands, environmental conditions, risk tolerances, and technical capabilities. Choosing the right control architecture is not merely a technological decision, but a strategic one that shapes the success, efficiency, and safety of unmanned operations across both civil and military domains.

CHAPTER 11

The Difference between (AI) and (ML)

8.1.1 What is (AI) in UAVs

Artificial Intelligence (AI) in targeting systems for Unmanned Aerial Vehicles (UAVs) has rapidly evolved into one of the most significant technological developments in modern warfare. As drones become more prevalent on the battlefield, their ability to operate autonomously, identify threats, and execute strikes without direct human input has transformed both strategic planning and tactical execution. AI-driven targeting has turned UAVs from remote-controlled surveillance assets into intelligent combatants capable of executing complex missions with speed, precision, and adaptability. Recent conflicts, particularly in Ukraine, the Middle East, and the South Caucasus, have served as proving grounds for these advancements, illustrating the growing importance of AI in modern military doctrine.

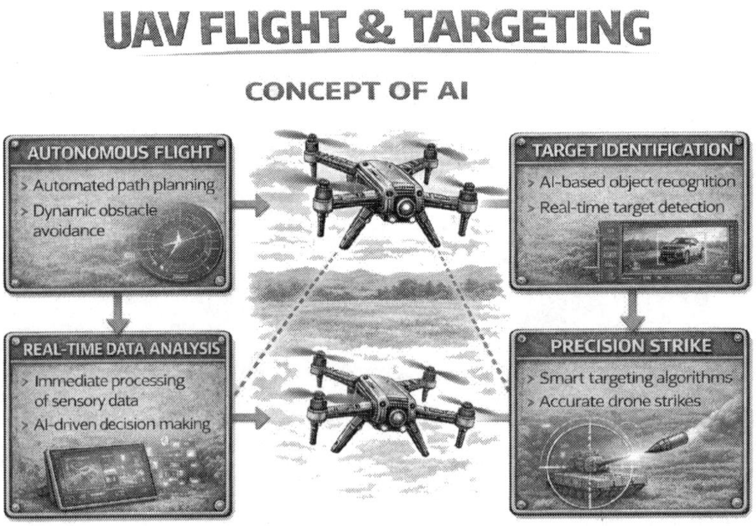

The core of UAV AI targeting lies in the ability to process vast amounts of real-time data from onboard sensors—including visual cameras, thermal images, radar systems, and acoustic detectors—and convert that information into actionable intelligence. Machine learning algorithms, trained on massive datasets comprising images of military vehicles, personnel, weapons systems, and infrastructure, allow drones to recognise and classify potential targets with increasing accuracy. Unlike traditional image processing, which relies on rule-based programming, AI systems improve over time as they are exposed to more data and environmental conditions. This self-improving capability makes AI particularly effective in the unpredictable and chaotic environments of modern combat zones, where the ability to quickly identify and respond to emerging threats is critical.

One of the significant milestones in the evolution of UAVs and the development of AI was the integration of GPS and inertial navigation systems (INS). GPS-enabled precise geolocation and route planning, allowing drones to navigate with high accuracy over long distances. However, because GPS signals can be jammed or spoofed, modern UAVs incorporate internal guidance systems, such as inertial navigation systems (INS) and terrain-contour matching, to maintain navigational integrity in GPS-denied environments. By blending multiple navigation sources, drones can autonomously reach and engage targets even under electronic warfare conditions.

Highly developed AI systems now rely heavily on autonomous visual targeting and object recognition. With the advent of high-resolution electro-optical (EO) and infrared (IR) sensors, UAVs can now conduct detailed visual surveillance in both day and night conditions. Combined with AI, these sensors enable drones to autonomously identify, track, and classify targets. Machine vision algorithms trained on vast datasets of vehicles, structures, and combatants allow UAVs to detect potential threats and prioritise targets in real-time. This visual targeting capability is especially crucial in urban or asymmetric warfare, where distinguishing between combatants and civilians is possible.

One of the most significant applications of AI in UAV targeting has been in loitering munitions, also known as "kamikaze drones." These systems are equipped with onboard AI that allows them to fly into a target area, scan for enemy assets, and autonomously decide when and where to strike. In the Russia-Ukraine conflict, both sides have made use of such drones, including AI-enhanced versions of the Lancet and Switchblade platforms. These drones are capable of identifying enemy tanks, artillery, radar installations, and even moving vehicles, engaging them with minimal human oversight. AI allows these drones to filter out non-combatants and decoys, making targeting more precise and reducing the likelihood of collateral damage—though, as with all autonomous systems, errors and misidentifications remain a significant concern.

In addition to offensive roles, AI-enabled targeting systems are playing an increasingly important part in defensive and reconnaissance missions. For example, drones equipped with AI-driven object detection can monitor large areas for hostile movements, identifying patterns that suggest an impending attack. By processing infrared, visual, and electromagnetic data, AI systems can detect camouflaged units, underground bunkers, or enemy equipment hidden among civilian infrastructure. Once a threat is identified, drones can autonomously shadow the target, maintain visual contact, or guide manned aircraft and artillery systems for engagement. In this way, AI significantly shortens the "sensor-to-shooter" timeline, which has traditionally relied on slow human interpretation and communication.

Another development in AI targeting is the implementation of multi-sensor fusion, where data from different types of sensors is combined to improve accuracy. For example, visual identification might be confirmed by heat signatures or radar profiles, reducing the likelihood of false positives. AI algorithms can also detect subtle patterns, such as vehicle track marks, heat exhaust anomalies, or irregular movement, that would be nearly impossible for a human operator to notice in real-time. These capabilities are particularly valuable in urban warfare or heavily forested areas, where line-of-sight is limited and the risk of misidentification is high.

AI also enables real-time adaptability. If a drone is approaching a heavily defended area, AI systems can alter flight paths, change targeting priorities, or even abort missions if the threat becomes too great. In swarm operations, multiple UAVs with AI-driven targeting systems can coordinate autonomously, sharing data and optimising target allocation. This form of distributed intelligence allows for highly flexible mission execution and makes it difficult for defenders to predict or counter drone behaviour. For instance, if one drone detects radar activity, it can signal others to divert or initiate an electronic warfare countermeasure.

Despite the obvious tactical advantages, the use of AI in targeting raises serious ethical and legal concerns. Autonomous systems challenge traditional rules of engagement and the principle of human accountability in combat. While most militaries insist on maintaining "human-in-the-loop" oversight—meaning a human operator must approve any lethal action—the trend toward greater autonomy is clear. In high-tempo operations or electronic warfare environments where communication is jammed or delayed, AI may be given increasing leeway to act independently. The potential for misuse or accidental targeting of civilians is a pressing issue, and there is an urgent need for international norms and treaties governing the deployment of such technologies.

In conclusion, the development and application of AI in UAV targeting have fundamentally altered the landscape of modern warfare. By enabling drones to identify, track, and engage targets autonomously, AI has vastly increased the speed, efficiency, and tactical versatility of aerial operations. These capabilities offer a clear battlefield advantage, especially in asymmetric conflicts or contested environments. However, they also introduce new challenges around control, accountability, and compliance with international humanitarian law. As AI technology continues to evolve, it will be critical for military planners, policymakers, and technologists to ensure that its application remains ethical, responsible, and under appropriate human oversight within the realm of total war and the overarching need to defeat the enemy.

8.1.2 What is (ML) in UAVs

Machine learning (ML) in UAV targeting refers to the use of data-driven algorithms that enable drones to identify, classify, track, and, in some cases, engage targets with increasing levels of autonomy and precision. Unlike traditional rule-based programming, where a drone operates based on pre-set instructions, machine learning allows the UAV to "learn" from data, adapt to new scenarios, and improve performance over time. In targeting, this means drones can be trained to recognise enemy vehicles, troops, weapons systems, or infrastructure based on visual, thermal, radar, or multi-spectral inputs. This capacity for pattern recognition and decision-making has transformed UAVs from passive tools into intelligent agents capable of executing dynamic, real-time combat operations. In recent warfare, particularly in Ukraine, Syria, and the South Caucasus, machine learning in targeting has been rapidly developed and widely deployed, becoming a cornerstone of modern drone-based combat tactics.

The development of machine learning in UAV targeting begins with data acquisition and model training. Huge datasets composed of satellite imagery, drone footage, and battlefield sensor data are used to train ML models—especially conventional neural networks (CNNs) and deep learning architectures—to recognise objects of military significance. These might include tanks, artillery pieces, radar dishes, enemy personnel, or even signs of recent vehicle movement. In addition to visual identification, ML models can be trained to fuse data from multiple sensors, such as combining infrared thermal signatures with electromagnetic signals to confirm the presence of hidden or camouflaged equipment. Through thousands or even millions of iterations, the machine learning model adjusts its internal parameters to improve its detection and classification accuracy. Once trained, these models are embedded in the drone's onboard computer or accessible via secure up-links, allowing real-time processing during flight.

In modern conflicts, the application of this technology has already shown immense battlefield value. For example, in the Russia-Ukraine war, both sides have used hobby and military drones modified with AI and machine learning models to detect enemy artillery positions or moving vehicles. Ukrainian forces, in particular, have used AI-enhanced drones for precise target acquisition, often coordinating with long-range artillery or HIMARS rocket systems. The drones fly into a target-rich environment, use machine learning algorithms to identify threats, and instantly relay coordinates with high confidence, dramatically accelerating the kill chain. In many cases, these ML systems can identify targets in low-light conditions, under camouflage netting, or even partially obscured by terrain—tasks that would be challenging or impossible for a human operator scanning footage manually.

Loitering munitions or "kamikaze drones" also benefit significantly from machine learning in targeting. These drones are designed to hover over an area until they identify a suitable target and then strike with lethal force. ML models allow these systems to autonomously scan the terrain, identify the most valuable or vulnerable target, and engage without waiting for human confirmation. In the 2020 Nagorno-Karabakh conflict, Azerbaijani forces used loitering drones equipped with ML-based target identification to devastating effect, eliminating tanks, armoured vehicles, and troop concentrations with surgical precision. These drones used computer vision techniques—trained via machine learning—to differentiate between legitimate military targets and civilian infrastructure, though with varying degrees of success.

Beyond combat roles, machine learning in targeting also supports broader battlefield intelligence. Drones equipped with ML algorithms can conduct persistent surveillance and detect patterns of enemy behaviour, such as the buildup of forces, movement of supply convoys, or emergence of new defensive positions. This predictive analysis enables commanders to make more informed strategic decisions and allocate resources effectively. In urban warfare or counterinsurgency operations, ML models can be trained to detect irregular movement patterns, such as vehicles deviating from expected traffic or the build up of forces seeking to engaging in offensive operations.

However, the application of machine learning in targeting is not without risks and challenges. One of the main concerns is the potential for false positives—when a drone misidentifies a civilian object as a military target. ML models are only as good as the data they are trained on, and in the chaotic and often ambiguous environments of modern warfare, errors can have catastrophic consequences. Adversarial actors may also attempt to fool ML systems by using decoys, camouflage, or data poisoning techniques to corrupt the drone's recognition capability. Moreover, when lethal action is driven by an algorithm, ethical and legal questions arise regarding accountability, proportionality, and compliance with the laws of armed conflict. To address these issues, many militarily should adopt a "human-in-the-loop" approach, where AI and ML systems assist in targeting but final approval is given by a human operator before being released to engage.

In summary, machine learning in UAV targeting is a trans-formative force in modern warfare, offering unmatched speed, precision, and adaptability in combat and reconnaissance roles. Its ability to analyse complex data, identify and track targets, and even predict future actions has given militarily new tools to dominate the battlefield. However, with this power comes responsibility, as the risk of misuse or unintended consequences remains significant. As the technology continues to evolve, it will be critical to balance the operational advantages of machine learning with robust oversight, ethical frameworks, and international cooperation to ensure its responsible deployment in future conflicts together with the use of AI.

8.1.3 The Difference between (AI) and (ML)

In the realm of unmanned aerial vehicles (UAVs), artificial intelligence (AI) and machine learning (ML) have become essential components in enhancing both flight autonomy and targeting capabilities. While these terms are often used interchangeably, they refer to distinct technological approaches with different functions, particularly in UAV systems. AI is a broader concept encompassing the simulation of human intelligence by machines—enabling them to perform tasks such as decision-making, reasoning, problem-solving, and adapting to new inputs. Machine learning, on the other hand, is a specific subset of AI that focuses on algorithms and statistical models that allow systems to learn from data, improve over time, and make predictions or classifications without being explicitly programmed. In UAVs, AI governs overall system intelligence, guiding how a drone navigates, makes decisions, or prioritises targets, while ML is the mechanism through which the drone processes data and improves its performance in specific tasks like object recognition or trajectory optimisation. Both technologies, though distinct, work together in modern UAVs to improve operational efficiency, reduce human intervention, and increase combat effectiveness, particularly in recent warfare.

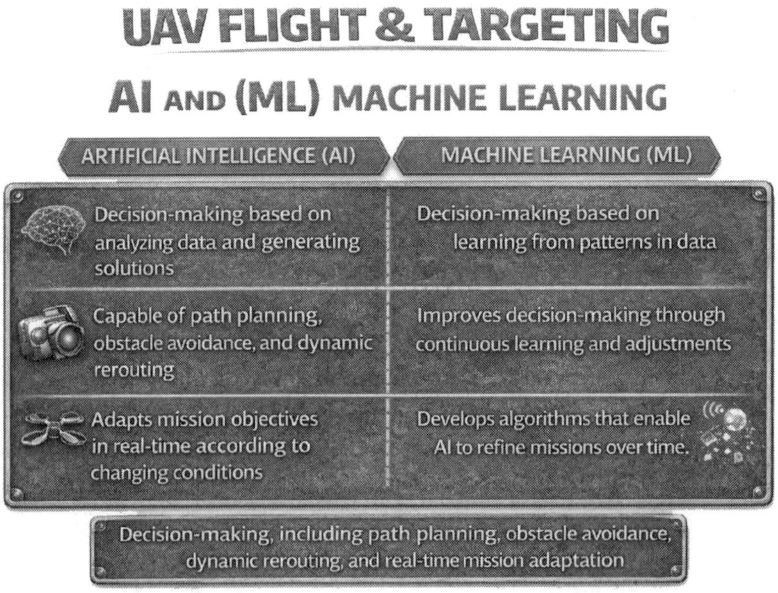

In flight operations, AI is responsible for overall autonomous decision-making, including path planning, obstacle avoidance, dynamic rerouting, and real-time mission adaptation. This includes systems that can assess changing environmental conditions such as weather, terrain, or enemy air defences, and alter their flight route or behaviour accordingly. AI allows UAVs to operate without constant human input, which is especially valuable in long-range missions or electronic warfare environments where communication may be compromised. For example, AI systems may determine that a UAV should fly at a lower altitude to avoid radar detection or divert its course due to an unexpected threat. These high-level decisions are based on rules, logic, and objectives programmed into the drone's software architecture.

Machine learning, meanwhile, enhances the UAV's low-level flight capabilities by enabling it to learn optimal control strategies or recognise patterns in flight data. ML can be used to train UAVs to stabilise in turbulent conditions, land in unfamiliar environments, or conserve battery life through efficient power management. Through reinforcement learning—an ML technique where systems learn optimal behaviour through trial and error—a UAV can teach itself how to best navigate a specific operational environment. ML algorithms may also detect anomalies in sensor readings or engine behaviour, predicting failures before they happen and improving system reliability. These ML-driven insights continuously feed into the AI decision-making system, helping it evolve its strategies based on real-time data.

The differences between AI and ML are also pronounced in UAV targeting systems. AI-driven targeting is about higher-order thinking: evaluating mission priorities, analysing the battlefield environment, coordinating with other assets, and determining whether to engage a target based on legal and ethical parameters. AI may take into account a wide range of variables, including the rules of engagement, civilian proximity, fuel reserves, and likelihood of mission success, before authorising a strike or recommending a course of action. This makes AI the 'brain' of the targeting process—guiding what to do and when to do it, often with human oversight.

ML, by contrast, is concerned with perception and recognition within the targeting process. ML models—especially those built with deep learning and computer vision—are trained to detect and classify targets using sensor inputs like high-resolution video, infrared imagery, or radar data. These models learn to distinguish between hostile and non-hostile objects by analysing vast datasets blabbed with various types of vehicles, personnel, buildings, and terrain. For instance, an ML model may be able to recognise a T-72 tank partially hidden in a forest, something that would be extremely difficult for a human operator or a rule-based AI system. Once a target is identified by ML, the AI component determines whether it is a viable target and, if so, how and when to engage.

Recent warfare has demonstrated the complementary application of both technologies. In the Russia-Ukraine war, drones such as Turkish Bayraktar TB2s and various commercial UAVs have reportedly used AI for autonomous navigation and targeting support, while leveraging ML to analyse surveillance footage and identify targets. Loitering munitions—also known as "kamikaze drones"—such as the Israeli Harop and U.S.-made Switchblade, employ ML to recognise specific target profiles and AI to decide engagement protocols, sometimes without requiring direct human command during critical moments. These capabilities allow for rapid strike decisions, minimised collateral damage, and greater operational flexibility.

As UAVs become increasingly autonomous, the integration of AI and ML is only expected to deepen. Swarm technologies—where groups of UAVs operate collectively—are also emerging as a battlefield asset, with AI managing the coordination and overall behaviour of the swarm, and ML enabling each unit to process local data and make micro-decisions. The blend of both allows swarms to navigate complex environments, divide tasks, and conduct simultaneous multi-target strikes with minimal human control.

In summary, while artificial intelligence and machine learning are often linked, their roles in UAV flight and targeting systems are distinct yet complementary. AI provides the overarching framework for autonomous decision-making and mission execution, while ML delivers the capacity to learn from data, recognise patterns, and optimise performance. Both technologies are now central to the development of next-generation drone warfare, enabling faster, smarter, and more autonomous operations on the battlefield. As conflicts become more data-driven and technologically complex, the effective integration of AI and ML into UAV systems will be a decisive factor in achieving aerial dominance.

8.1.4 UAV Levels of Flight Autonomy

Level 0 - No Drone Autonomy

Drones that operate at level 0 autonomy need a pilot in control of the drone 100% of the time. Without a pilot, the drone will crash. Level 0 drones lack the ability to understand or respond to obstacles. They are typically used for recreational purposes such as racing.

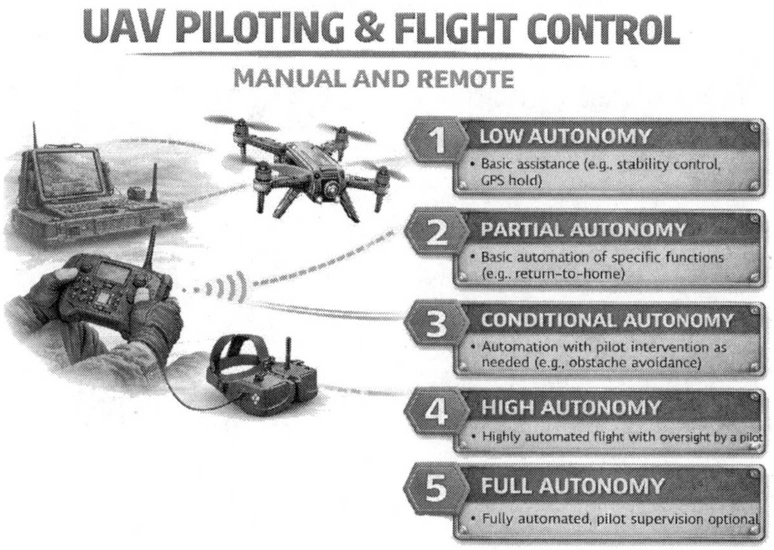

Level 1 - Low Drone Autonomy

Level 1 drones operate with low autonomy, which can account for the spatial limitations of an environment, such as walls or ceilings, allowing them to safely operate within an enclosed space. These drones can remain in the air without a pilot but must stay in the pilot's Visual Line of Sight (VLOS). Level 1 drones are usually toy drones with six-axis gyro sensors to aid in stabilisation and sometimes include a "return home" button on the controller, which returns the drone to its launch point when pressed.

Level 2 - Partial Drone Autonomy

In level 2 drones, autonomous systems are more advanced, but the pilot is still in complete control. Although these drones use a combination of sensors, accelerometers, and GPS receivers, a pilot still operates its movements and receives warnings when the drone approaches an obstacle so the pilot can steer it to safety. Level 2 drones cannot detect and avoid or navigate, but they can detect and warn. Like level 0 and level 1 drones, level 2 drones and operate in the pilot's VLOS.

Level 3 - Conditional Drone Autonomy

Level 3 drones operate with a high level of autonomy in that the pilot is no longer flying the drone but is onsite as a backup in case of an emergency. Most drone delivery operators currently function at a level 3 autonomy, where pilots are in place only for emergencies. These drones are equipped with basic detect and avoid systems that use radio and frequency sensors, which help them avoid obstacles such as buildings or poles in the flight path. Level 3 drones can also find safe spaces to land autonomously based on their awareness of the environment.

It's important to note that level 3 drones in the drone delivery market face regulatory hurdles driven by existing aviation laws that focus on a pilot in the cockpit of an aircraft making decisions. Some commercial delivery drones operate under a Part 107 waiver from the FAA, but there are strict limitations to ensure safety. For example, under Part 107, drones may not fly BLVOS, they may not fly over people or roads, humans must be present on-site, and one pilot can operate only one drone at a time.

Level 4 - High Drone Autonomy

Level 4 drones have a high level of autonomy in that they use advanced detect and avoid systems and detect and navigate systems without pilot action. They can freely explore GPS-free and GPS-denied environments, navigate harsh conditions, and identify people in need without needing a pilot on-site, as pilots can monitor level 4 drones remotely. As FAA regulations loosen, one pilot will be able to manage many BVLOS drones simultaneously. For example, Zipline has scaled up to 24 drones in one fleet under the command of a single operator in Africa. Eventually, an unmanned traffic management system (UTM) will manage drone operations rather than pilots.

Exyn's drones use GPS-free navigation to explore complex spaces without a pilot as a backup freely. These drones build maps of their surroundings while tracking their movement through the environment, leading to self-reliant drones exploring independently without human interaction during flight. This is a significant step up from the level 3 drones, which need a human operator or driver to be present to take control of the system at any time.

Keep in mind, Exyn's drones reached level 4 autonomy by exploring human-free environments such as mines and caves; they are not operating above roads or humans, so regulations are not as strict. Drone delivery operators, on the other hand, will need BVLOS flight clearance and go through the Part 135 certification process at the FAA to operate drones over people and roads. While this is no easy task, AI and computer vision will address these hurdles allowing for safe and predictable drone operations.

Level 5 - Full Drone Autonomy

At the time of writing this book in 2025, Level 5 drones were still considered by many to be a distant goal.

These systems are expected to operate in any environment, under all conditions, and with complete independence from human intervention. Achieving Level 5 autonomy requires drone navigation performance comparable to—or exceeding—that of aircraft reliant on external guidance systems. Currently, no Level 5 drones are in mass production due to several challenges: an evolving regulatory framework, the need for advanced airspace management systems capable of coordinating both unmanned and crewed aircraft, and the immense data-handling requirements associated with fully autonomous flight.

As the autonomous drone sector continues to expand at remarkable speed, AI developers face mounting pressure to design, test, and commercialise solutions that enhance both safety and reliability. With drones, the margin for error is effectively zero; any failure has the potential to cause serious injury or death to humans or animals. The integration of AI and computer vision will be central to mitigating these risks. It appears only a matter of time before drones become consistently safe, fully compliant with regulatory expectations, and ready for large-scale deployment.

However, outside of the civil regulatory framework, during 2024-2025 the author of this book, Peter Savage, observed factual evidence indicating that a fully autonomous fixed-wing drone already exist. Developed from a highly modified fixed-wing hobby kits, this platform integrated next-generation solid-state data storage—approximately the size of a fifty-cent coin yet capable of holding four, eight, or even more terabytes of data storage—with miniaturised four- to eight-core processing boards similar to the Raspberry Pi 5 or advanced Arduino-class motherboards. This technological fusion made true end-to-end autonomy possible for specifically targeted mission profiles.

The resulting systems formed the first generation of drones capable of identifying terrain along predefined routes, recognising threats, and dynamically adjusting their trajectory. The substantial onboard storage enabled real-time machine-learning processes, allowing the aircraft to conduct autonomous flight while communicating and navigating through mesh networks or GPS-enabled (SATCOM) environments. While early prototypes were initially deployed for surveillance, it soon became clear that they offered far greater value as long-range hard-kill platforms—particularly in missions launched from Ukraine deep into Russian territory.

This emerging high-end capability has fundamentally reshaped our understanding of autonomous flight. It has only become possible through the convergence of several key technologies: high-capacity miniature storage drives, low-power multi-core processors, and highly efficient high-throughput data-processing architectures. The true key to autonomy lies in the system's ability to access and process vast quantities of data at speed, enabling real-time terrain, target, and environmental identification while executing complex mission directives with precision.

CHAPTER 12

UAV Evasive Flight Patterns, Stealth Materials

9.1.1 Evasive Quad-Copter Flight Patterns

In the context of modern warfare, quad-copter UAVs—commonly used for reconnaissance, surveillance, targeting, and even direct strikes—have become highly valuable assets on the battlefield. However, their increased presence has made them primary targets for enemy countermeasures, particularly small arms fire, electronic jamming, and even net-based or kinetic drone counter-systems. As a result, the development and implementation of evasive flight patterns for quad-copters have become crucial for operational survivability. These evasive manoeuvres are designed to reduce the UAV's visibility, predictability, and vulnerability during missions, especially in contested environments where adversaries are deploying both kinetic and non-kinetic anti-drone technologies.

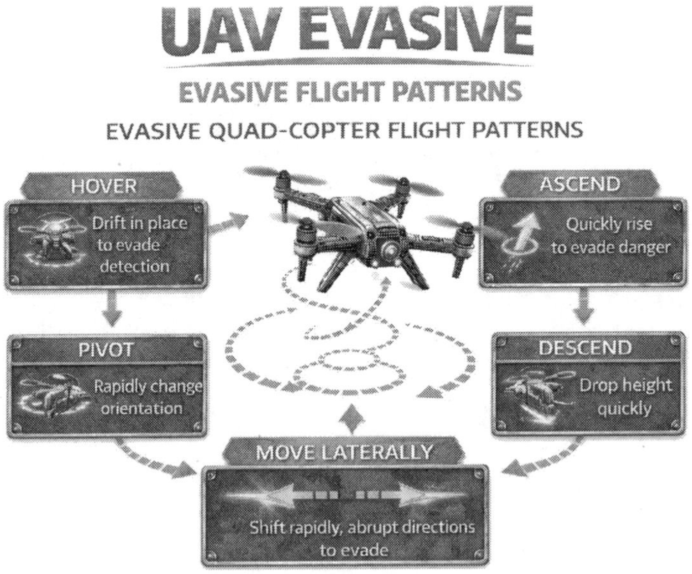

By using pre-programmed evasive routines, machine learning algorithms, or real-time operator adjustments, quad-copters can increase their chances of mission success while avoiding destruction or capture. Quad-copters are particularly well-suited to perform agile and unpredictable flight manoeuvre due to their unique flight dynamics and control systems. Unlike fixed-wing drones, which rely on forward motion and aerodynamic lift, quad-copters use multiple rotors for vertical takeoff and omni-directional movement. This enables them to hover, pivot, ascend, descend, and move laterally in rapid, abrupt directions. In warfare, this translates into the ability to suddenly change altitude, jink from side to side, or perform zigzag movements while approaching a target or retreating from a hostile area. These evasive patterns disrupt the aiming calculations of enemy shooters and reduce the effectiveness of anti-drone systems that rely on predicting flight trajectories. For instance, when quad-copters are used to drop munitions or observe front-line movements, a sudden zigzag or drop in altitude before engaging can make it harder for adversaries to lock on or shoot down the UAV.

Evasive manoeuvres can be manually controlled by skilled drone pilots or autonomously executed using artificial intelligence and onboard sensors. Manual evasive flying is common among experienced operators who anticipate threats and rely on real-time feedback from the UAV's camera feed. However, in high-threat environments, latency or signal disruption can hinder manual responses. To address this, militarily and insurgent groups alike are increasingly relying on automated evasive flight software, which allows the drone to detect incoming threats—like radar locks, jamming signals, or muzzle flashes—and automatically alter its flight path. Some advanced systems incorporate machine learning algorithms that analyse historical engagement patterns and learn which evasive movements are most effective against specific types of threats, such as small-arms fire versus jamming zones. Over time, the drone refines its behaviour and adapts its evasion tactics to match the evolving threat landscape.

In electronic warfare environments where GPS jamming or signal spoofing is prevalent, evasive flight patterns become part of broader survivability routines. For instance, when a quad-copter detects GPS interference, it may automatically shift into a non-GPS-based inertial navigation mode, while simultaneously adjusting its altitude and course to exit the jamming zone. In such scenarios, the quad-copter may follow erratic, curved escape routes rather than linear trajectories to prevent enemy tracking or signal interception. Some drones are even programmed to perform "evasive hovering," where the UAV randomly shifts its position slightly while stationary to avoid appearing as a static target. These subtle, unpredictable movements increase drone survivability, especially during prolonged over-watch or surveillance missions near front-lines or high-value targets.

Another key component of evasive flight is terrain masking and vertical manoeuvring. By flying low to the ground or staying close to buildings, trees, or other obstacles, quad-copters can reduce their visibility to both human observers and radar systems. This approach, sometimes called "map-of-the-earth" flying, takes advantage of the UAV's small size and manoeuvrability to avoid detection or lock-on. When paired with vertical climbs or drops—such as rising above treetops only to immediately descend behind a building—drones can obscure their flight path and evade tracking systems. In urban warfare, where line-of-sight is often obstructed, these techniques are particularly useful for avoiding drone interception nets, gunfire, and signal triangulation.

Swarm-based tactics have also introduced a new form of evasive manoeuvring. In swarm deployments, multiple quad-copters are flown together with randomised or semi-synchrony- flight paths. The intent is to overwhelm enemy defences with multiple targets, forcing adversaries to either divide their attention or expend resources inefficiently. Individual drones in the swarm may perform unpredictable movements, sudden direction changes, or time-delayed dispersal's, making it difficult to determine which drone carries the payload or serves as the primary reconnaissance unit. Even if several drones are intercepted, the overall mission can still succeed, demonstrating the tactical advantage of swarm-based evasion.

In summary, evasive flight patterns in quad-copter UAV warfare are essential for enhancing survivability and mission effectiveness in increasingly hostile and technology-saturated battlefields. By combining manual agility, AI-based automation, terrain awareness, and swarm dynamics, modern quad-copters can evade a wide range of threats—from bullets and missiles to radar and jammers. As counter-drone technologies continue to evolve, so too will the complexity and sophistication of evasive flight strategies, ensuring that quad-copters remain a viable and potent tool in both conventional and irregular warfare settings.

9.1.2 Evasive Fixed-Wing Flight Patterns

Fixed-wing unmanned aerial vehicles (UAVs) have become an indispensable component of modern military operations, offering long-range surveillance, reconnaissance, and strike capabilities. Unlike quad-copters, which are prised for their agility and vertical manoeuvrability, fixed-wing UAVs excel in endurance, altitude, and speed, often operating across vast areas for extended durations. However, their reliance on aerodynamic lift and continuous forward motion limits their ability to hover or make abrupt changes in direction. This inherent limitation makes them more vulnerable to detection and interception, particularly in contested environments where air defences and counter-UAV measures are actively employed. As a result, the development and implementation of evasive flight patterns tailored to fixed-wing UAVs have become critical for maintaining survivability, mission continuity, and overall operational effectiveness in warfare.

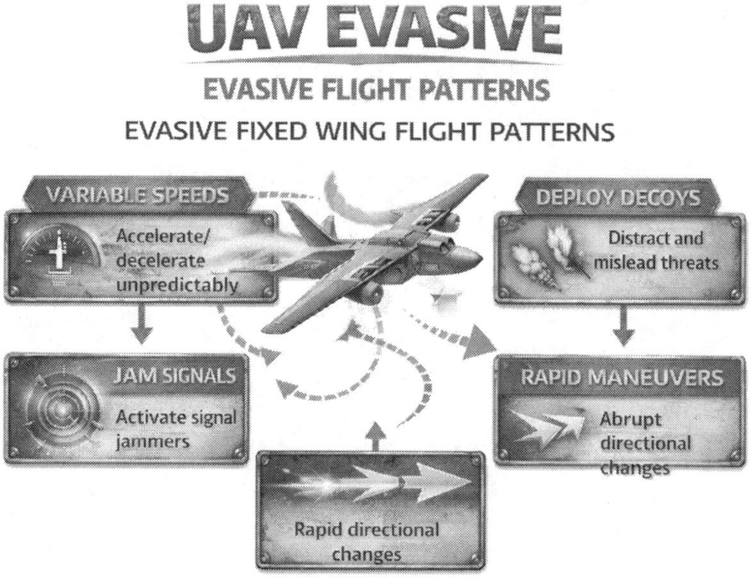

Fixed-wing UAV evasive maneuvers are largely dependent on exploiting the aircraft's speed, altitude, and aerodynamic capabilities. These evasive flight patterns are designed to disrupt enemy targeting, reduce predictability, and increase the difficulty of successful engagement by anti-aircraft systems, radar-guided weapons, or hostile aircraft. One common tactic is the use of high-speed passes combined with altitude variation. By rapidly ascending or descending in response to threats, fixed-wing drones can avoid surface-to-air missile (SAM) lock-ons and frustrate radar tracking systems. This "terrain skimming" or "terrain masking" technique involves flying at low altitudes where the UAV can use natural terrain features—like hills, valleys, and buildings—to obscure itself from radar detection and visual observation. Once the threat is passed or the mission is complete, the UAV may then climb back to cruising altitude for safe return or to reestablish communications.

Another core evasive maneuvers involves serpentine or zigzag flight paths. Instead of flying in a straight line, which is easier for enemy radar and anti-aircraft systems to predict, fixed-wing UAVs employ erratic lateral deviations within their flight corridors. These deviations are often pre-programmed into their navigation systems or triggered by onboard threat detection systems. This unpredictability complicates targeting solutions for guided munitions and increases the time window needed for an adversary to respond. These flight paths may be enhanced with variable speeds—accelerating or decelerating at unpredictable intervals—further disrupting enemy engagement timing. In high-risk zones, particularly near known SAM installations or man-portable air-defence systems (MANPADS), these techniques significantly increase the drone's chance of evading a successful hit.

Modern fixed-wing UAVs also utilise electronic countermeasures (ECM) as part of their evasive behaviour. When integrated with evasive flight programming, ECM systems can detect radar emissions or missile launches and automatically initiate a series of defensive actions. This may include deploying decoys, activating signal jammers, or initiating a rapid climb or dive to escape the engagement zone. These systems are often paired with onboard AI capable of recognising patterns associated with radar lock-on or enemy interception attempts. Once a threat is detected, the AI system may autonomously alter the UAV's route or switch into an evasive flight profile to maximise survivability without requiring operator input—a valuable feature in communications-degraded environments where manual intervention may be delayed or impossible.

In environments where GPS jamming or spoofing is employed, fixed-wing UAVs may initiate predefined evasive protocols, relying on inertial navigation systems (INS) and terrain contour matching to maintain flight integrity. These drones can be programmed to avoid known electronic warfare hot-spots, flying along curved or staggered paths to make it harder for enemies to calculate intercept vectors. Some advanced UAVs are capable of "way-point randomisation," which means that even if the start and end points of a mission are known, the path taken changes in each mission to minimise pattern recognition and reduce the risk of ambush by enemy forces. These changes are often informed by machine learning systems that process previous flight data and identify patterns in adversary behaviour or defence posturing.

Another key element in fixed-wing evasive flight is the use of high-altitude operations. Operating at altitudes beyond the reach of most small arms and short-range air defense systems, high-flying UAVs can perform surveillance and reconnaissance with relative safety. However, when missions bring them into lower altitude bands—especially during strike operations or close-range intelligence gathering—evasive patterns become crucial. This is when integration with multi-sensor detection systems becomes invaluable, allowing drones to recognise heat signatures, radar sweeps, or missile launches, and respond in real-time with appropriate maneuvers or electronic deception.

Recent conflicts, including those in Ukraine, Syria, and the Caucasus, have demonstrated the increasing sophistication of fixed-wing UAV evasive capabilities. Both state and non-state actors have used these tactics to keep drones operational in dense anti-air environments. Turkish Bayraktar TB2s, for example, have demonstrated evasive behaviour when under threat, using terrain-following techniques and route variation to avoid Russian-made air defence systems. Likewise, Russian Orion drones and U.S.-built MQ-9 Reapers employ similar strategies when flying in contested airspace, often adjusting flight paths dynamically in response to enemy activity.

In summary, evasive flight patterns in fixed-wing UAV warfare are a critical component of survivability and mission success. While these aircraft lack the agility of multi-rotor platforms, they compensate with speed, altitude, and strategic manoeuvring supported by AI and electronic warfare systems. Through techniques like terrain masking, serpentine routing, altitude modulation, and adaptive flight planning, fixed-wing drones are increasingly able to survive in hostile environments and carry out complex, high-risk missions. As adversaries continue to advance their counter-drone technologies, the importance of evolving and refining these evasive strategies will only grow.

9.1.3 Stealth Composite Materials, Structure

The construction of unmanned aerial vehicles (UAVs) has undergone a dramatic transformation over the past few decades, largely driven by advances in materials science and engineering. At the heart of this transformation is the widespread adoption of composite materials, which have revolutionised how UAV air-frames are designed, manufactured, and utilised on the battlefield. Traditional metallic materials such as aluminium and titanium once dominated aerospace construction due to their relative lightness and strength. However, the increasing demand for lighter, more durable, and lower-signature platforms has shifted attention toward non-metallic alternatives—chief among them being composite materials such as carbon fibre-reinforced polymers (CFRPs), fibreglass-reinforced plastics (FRPs), and various hybrid laminate structures. These composites have not only improved the performance characteristics of UAVs but have also broadened the strategic and operational possibilities of unmanned systems in both conventional and asymmetric warfare.

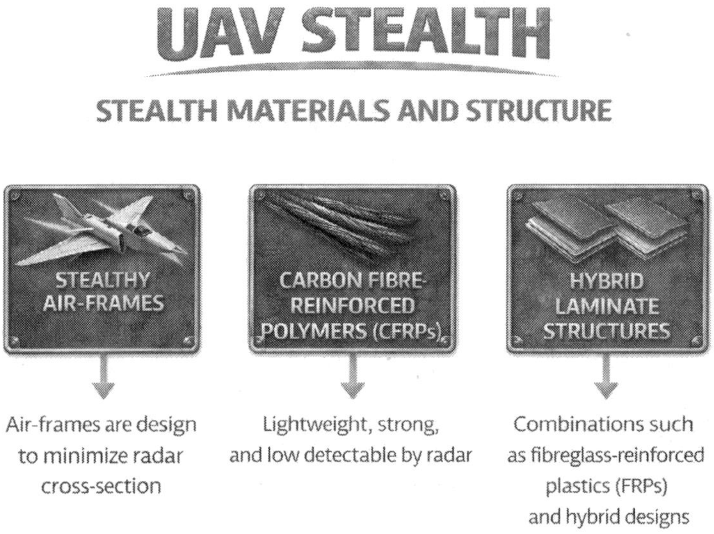

One of the most significant advantages offered by composite materials is their superior strength-to-weight ratio. This property is particularly crucial in UAV design, where weight savings directly correlate with improvements in endurance, range, and payload capacity. A lighter air-frame allows the UAV to carry more surveillance equipment, sensors, or even munitions without compromising flight time. Additionally, lower air-frame weight reduces fuel consumption or battery drain, thereby extending mission duration and reducing the logistical burden of resupply or battery swapping. Carbon fibre composites are especially valued for these characteristics. With tensile strength significantly higher than steel and a fraction of its weight, carbon fibre allows UAVs to achieve higher performance metrics while maintaining structural integrity under the stresses of flight, particularly at high speeds or during complex maneuvers.

Fibreglass, another widely used composite material, offers a more cost-effective alternative with good mechanical strength, impact resistance, and a degree of flexibility that is beneficial in absorbing vibrations and minor shocks. It is often used in smaller or commercial-grade UAVs that require cost-efficient mass production while still benefiting from the mechanical advantages of composite construction. Hybrid laminates —combinations of carbon fibre, aramid (such as Kevlar), and fibreglass—are also gaining traction for military UAVs that must balance strength, flexibility, and damage tolerance. These multi-material layups provide customisation properties, allowing engineers to tailor sections of the air-frame for specific performance needs, such as reinforcing wing roots for load-bearing, or incorporating Kevlar near vulnerable components for ballistic protection.

In addition to structural performance, composite materials contribute significantly to UAV survivability through corrosion resistance and reduced detectability. Unlike metals, composites do not oxidise or degrade when exposed to moisture, salt, or various environmental conditions—an essential trait for UAVs operating in maritime or tropical environments. This resistance to corrosion reduces maintenance cycles and extends operational life, especially for drones deployed in forward or remote areas with limited support infrastructure. From a stealth perspective, the non-metallic nature of composites offers another major advantage: a reduced radar cross-section (RCS). Metal structures inherently reflect electromagnetic waves, making them more visible to radar detection. Composite materials, particularly those embedded with radar-absorbent properties, help minimise this reflectivity. Their molecular structure can be tailored to absorb or scatter radar energy rather than reflecting it directly back to the source, thereby enhancing the UAV's stealth capabilities and making it more suitable for missions in contested or anti-access/area denial (A2/AD) environments.

The manufacturing of composite UAV structures requires specialized fabrication techniques, many of which have matured in parallel with the evolution of UAV design. One of the most widely used methods is the pre-impregnated (pre-preg) layup process. In this approach, carbon or glass fibre fabrics are pre-saturated with a thermo-setting resin and stored under refrigerated conditions until use. During fabrication, these pre-pregs are carefully laid into moulds layer by layer, with fibres oriented in specific directions to achieve the desired mechanical properties. The layup is then vacuum-sealed and cured in an autoclave under controlled heat and pressure. This method ensures high strength and minimal voids, making it ideal for critical structural components like wings, fuselage shells, and control surfaces.

For larger UAVs or those requiring faster production cycles, vacuum-assisted resin transfer moulding (VARTM) is often employed. This technique involves placing dry fibre reinforcements into a mold, then using vacuum pressure to draw liquid resin through the fibres. Once the resin saturates the reinforcement, the entire assembly is cured—typically without the need for an autoclave. VARTM is particularly suitable for medium-scale production of composite UAV components and offers a balance between performance and cost. It also allows for the incorporation of core materials, such as foam or honeycomb structures, which enhance stiffness and strength without adding excessive weight.

The advent of additive manufacturing has further expanded the possibilities for UAV construction using composites. 3D printing technologies capable of handling fibre-reinforced filaments or resin-based composites now allow for rapid prototyping and even field-level production of air-frame parts. These technologies support on-demand customisation, enabling forces to tailor UAV components to specific mission parameters or environmental conditions. For instance, different nose cone shapes or payload housings can be printed to reduce RCS or improve aerodynamic stability, and damaged components can be replaced quickly in-theatre without requiring extensive logistics chains. While 3D printed composites may not yet match the mechanical properties of autoclave-cured pre-pregs, advances in continuous fibre printing and thermoplastic composites are rapidly narrowing that gap, making them increasingly viable for military applications.

Another important development in composite UAV construction is the integration of multi-functional materials—composites that not only serve structural roles but also provide additional capabilities. For example, carbon fibre laminates embedded with conductive materials can double as antennas, reducing the need for protruding structures that compromise stealth. Similarly, composites can be layered with sensors or thermal insulation, contributing to the drone's situational awareness or infrared signature reduction. Research into smart composites is also promising, with materials that can self-monitor for cracks or stress accumulation, and in some cases, self-heal using embedded micro-capsules or thermally activated resins. These innovations represent the convergence of materials science, structural engineering, and mission-specific design philosophies that will define the next generation of UAVs.

In conclusion, the rise of composite materials in UAV construction has been a cornerstone of their transformation from basic reconnaissance tools into highly capable, survivable, and adaptable combat platforms. The exceptional strength-to-weight ratios, corrosion resistance, and stealth-friendly properties of composites have enabled UAVs to operate longer, carry more, and evade detection more effectively than ever before. As manufacturing techniques like pre-preg layups, vacuum infusion, and 3D printing continue to advance, and as new multi-functional materials are developed, the strategic importance of composites in UAV warfare will only grow. These materials not only improve the technical performance of UAVs but also expand their tactical and operational roles across the full spectrum of modern military engagements. The evolution of UAVs, therefore, is inextricably tied to the continued innovation and application of advanced composite structures.

9.1.4 Signature Suppression, Radar-Absorbent (RAM)

Building upon the structural and material foundation provided by composite construction, the next critical area of advancement in unmanned aerial vehicle (UAV) technology lies in the development and application of (RAM) stealth systems. In the age of networked warfare, where battlefield survivability is directly tied to electronic visibility, reducing a UAV's detectability across multiple sensor spectrum's is paramount. Stealth in UAVs is no longer confined to reducing radar cross-section (RCS); instead, it encompasses a broad range of strategies and technologies aimed at minimising emissions and reflections in the radar, infrared (IR), visual, and acoustic domains. This form of multi-spectral signature management is essential for ensuring UAVs can penetrate defended airspace, perform covert surveillance, or conduct electronic and kinetic strikes without triggering early warning systems.

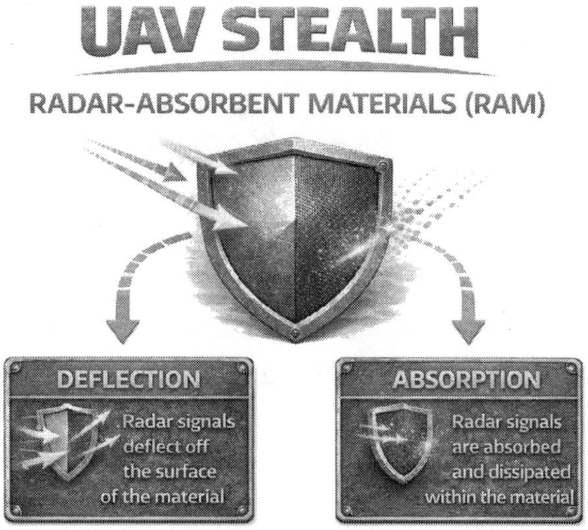

At the core of radar signature suppression is the use of radar-absorbent materials (RAM) and strategic shaping of air-frames. RAM technologies involve coatings and materials that absorb incoming electromagnetic waves instead of reflecting them back to the radar source. This reduces the radar cross-section and makes the UAV appear smaller—or even invisible—on radar scopes. RAM is commonly made from materials like carbonyl iron or ferrite-based compounds embedded in paint, foam, or polymer layers applied to the UAV's exterior. These materials are designed to absorb specific radar frequencies, converting the energy into heat rather than reflecting it. Some UAV designs take this a step further by embedding radar-absorbing materials directly into the composite structure itself, creating air-frames that are not only strong and lightweight but also inherently radar resistant. For example, certain carbon fibre-reinforced polymers can be doped with conductive fillers that disrupt radar wave propagation. The combination of RAM and advanced composites enables UAVs to achieve low observability without relying solely on surface coatings that can degrade or be damaged in the field.

Complementing the use of RAM is the deliberate shaping of UAV air-frames to deflect radar waves away from their source. Known as radar-deflective geometry, this approach relies on angled surfaces, faceted shapes, and smooth, continuous contours that minimise the likelihood of a radar wave being reflected directly back to a receiver. Unlike traditional aircraft that often feature perpendicular joints and protruding elements, stealth UAVs use blended wing-body designs, embedded engine nacelles, and recessed control surfaces to create an unbroken profile. These aerodynamic features serve the dual purpose of enhancing flight efficiency and disrupting radar detection. Some advanced UAVs also feature edge alignment and "V"-shaped trailing edges to scatter radar energy in multiple directions.

This principle was famously applied in the design of manned stealth aircraft such as the F-117 Nighthawk and B-2 Spirit, where faceted or blended air-frames were used to scatter radar waves and minimise detection. Over time, these design philosophies have been miniaturised and integrated into advanced unmanned aerial vehicles (UAVs) like the RQ-170 Sentinel and the Kratos XQ-58A Valkyrie, which feature smooth contours, internal weapon bays, and radar-deflective geometries to reduce their radar cross-section. However, the influence of stealth shaping and signature management is not limited to high-end military platforms. In the ongoing war in Ukraine, even modified hobby drones have adopted rudimentary stealth concepts. Civilian quad-copters, for instance, are often painted in matte, low-reflectivity coatings, constructed in carbon fibre or covered in makeshift camouflage to reduce their visibility to both radar and optical systems. Some hobby drones are also modified to present smaller profiles or carry minimal reflective material, which, while initially crude compared to military-grade stealth technologies, more advanced techniques are now being used to lower detectability in battlefield conditions. These adaptations reflect a broader trend where stealth principles are being democratised and applied across a wide spectrum of UAV platforms, from sophisticated strike drones to improvised commercial quad-copters used in modern asymmetric warfare.

Infrared (IR) signature suppression is another vital component of UAV stealth. Most military sensors today include IR detection capabilities that track heat emitted by engines, electronics, or even friction from high-speed flight. Heat-seeking missiles and infrared tracking systems pose a particular threat to UAVs operating at low altitudes or loitering near high-value targets. To counter this, UAV designers incorporate several heat-reduction strategies. One method is the use of shielded or diverted exhaust systems that mix hot gases with cool ambient air before release, thereby lowering the UAV's thermal footprint. Some designs relocate exhaust vents to the top of the air-frame, away from line-of-sight detection from the ground. Additionally, heat sinks, ceramic tiles, and phase-change materials can be used to absorb and dissipate engine heat gradually, making the UAV harder to detect with thermal imaging equipment.

Electric propulsion systems offer a stealth advantage in this domain as well. Battery-powered UAVs generate significantly less heat than those with combustion engines, making them inherently less visible in the infrared spectrum. This characteristic has made electric UAVs especially valuable in tactical reconnaissance and special operations where close-range observation without detection is critical. Moreover, thermal-insulating composites and reflective surface coatings can be applied to the UAV's exterior to mask the internal heat profile and harmonise it with the ambient environment. These coatings are often tuned to specific wavelengths to avoid detection by common IR sensors used in military operations.

Visual stealth remains a deceptively simple yet important aspect of UAV low-observability. In daylight operations, UAVs are susceptible to detection by ground observers and optical sensors, particularly if they leave visible trails or operate at low altitudes. As a result, many military UAVs are painted with matte, non-reflective coatings that reduce glare and eliminate reflections from sunlight. Camouflage patterns—whether digital, geometric, or gradient-based—are also tailored to match the expected operational environment, whether it be urban, desert, forested, or maritime. Visual stealth strategies are especially effective when paired with flight profile management, such as flying at dusk, dawn, or in overcast conditions where UAVs are harder to spot against the sky.

Emerging technologies are pushing visual stealth even further through the use of adaptive camouflage systems. These "smart skins" can adjust colour, texture, or even luminosity in response to environmental cues. Inspired by biological systems like cephalopods, adaptive camouflage may eventually allow UAVs to blend dynamically into their surroundings in real-time. While still largely experimental, developments in electro-chromic and thermo-chromic materials suggest that UAVs could, in the near future, possess skins that actively respond to light, heat, or electromagnetic fields. These innovations hold promise not only for visual blending but also for reducing IR and radar signatures through changes in emissivity and reflectivity.

Acoustic signature reduction is another increasingly important stealth consideration, particularly for UAVs operating near enemy positions or over populated areas. Propeller noise, engine whine, and vibration all contribute to the overall sound profile of a UAV. To reduce this, engineers employ low-noise propellers with specialised blade geometries, variable pitch mechanisms, and dampening mounts for motors and electronics. Some UAVs are designed to operate at frequencies less perceptible to the human ear or to mask their acoustic footprint by flying at higher altitudes. Again, electric UAVs tend to fare better in this category, as they produce minimal mechanical noise compared to combustion-powered counterparts. In swarm operations, noise suppression can be essential in delaying the enemy's realisation that drones are overhead, which can be crucial in delaying a defensive reaction or preserving the element of surprise.

The convergence of stealth strategies across these multiple spectrum's is giving rise to a new generation of UAVs that are difficult to detect and even harder to counter. A growing area of research is the integration of multi-spectral stealth materials—engineered to simultaneously minimise detectability across radar, IR, visual, and acoustic domains. These materials combine RAM properties, thermal regulation, visual camouflage, and noise dampening into single-layer or multi-layer skins. For example, nano-composite coatings are being developed that use graphene, carbon nanotubes, and metal oxides to selectively absorb and scatter electromagnetic energy while also providing thermal conductivity and UV protection. These coatings not only improve stealth but also enhance durability and weather resistance.

Another promising innovation is the use of meta-materials—engineered structures that interact with electromagnetic waves in unconventional ways. Unlike traditional materials, meta-materials can be designed to bend, absorb, or redirect radar waves regardless of the material's base composition. UAVs equipped with meta-material coatings or structural panels may be able to achieve unprecedented levels of radar stealth, especially when combined with shape-based RCS reduction. Some theoretical meta-material configurations have even demonstrated potential for cloaking effects at certain frequencies, although practical implementation remains at an early stage.

In summary, the integration of stealth technologies into UAVs has become a multidimensional and highly technical endeavour, aimed at minimising detectability across the full electromagnetic and acoustic spectrum. From the use of radar-absorbent materials and deflective shaping to infrared suppression, visual camouflage, and noise reduction, modern UAV design is defined as much by its ability to evade detection as by its payload or endurance. As materials science, computational design, and adaptive technologies continue to evolve, the future of UAV stealth lies in systems that are not only structurally optimised but also dynamically responsive to the changing environments and threats they encounter. In this way, stealth has become not just a defensive measure but a proactive enabler of operational dominance in the increasingly contested domains of modern warfare.

9.1.5 UAV Radar Reflectivity Science, Principles

Central to the stealth discussion in unmanned aerial vehicle (UAV) design is a detailed understanding of radar reflectivity science and the deliberate management of radar cross-section (RCS). Radar cross-section is a quantitative measure of how detectable an object is by radar, and its manipulation lies at the heart of all stealth technologies. The ability of a UAV to avoid detection or tracking by radar is not a singular function of its size but a complex interaction between electromagnetic waves and the geometry, material properties, and surface characteristics of the UAV. As radar detection systems have grown more sophisticated and widespread—particularly in high-threat environments and contested airspace—UAV designers have focused considerable effort on understanding and minimising the RCS through a combination of structural design, material engineering, and electronic emission management.

The physics of radar detection begins with the propagation of radio frequency (RF) electromagnetic waves emitted from a radar transmitter. These waves travel outward until they encounter an object, at which point they may be absorbed, (As outlined in the RAM section, 9.1.4), refracted, scattered, or reflected. It is the energy that is reflected back toward the radar receiver that determines whether an object is detected and, if so, how it is interpreted in terms of location, size, speed, and movement. The magnitude and direction of this reflected energy depend on several key variables, including the shape and orientation of the object, the materials it is composed of, and the texture and smoothness of its surface. Radar systems typically operate within a range of frequencies—from L-band (1-2 GHz) to X-band (8-12 GHz)—and the specific radar frequency influences how waves interact with different materials and surfaces. High-frequency radar provides finer resolution but is more susceptible to attenuation, while lower frequencies can penetrate foliage or detect low-RCS objects but with less spatial accuracy.

One of the most important factors influencing RCS is the shape of the UAV. When radar waves strike flat, perpendicular surfaces, a significant portion of the energy is reflected directly back toward the radar source, creating a strong return signal. To minimise this, stealth UAVs are designed with faceted or curved surfaces that deflect radar waves away at oblique angles, reducing the energy that is reflected back to the receiver. Faceted designs, such as those used in early stealth aircraft like the F-117 Nighthawk, consist of flat panels arranged in such a way that radar waves are bounced off in multiple directions. More modern UAVs employ blended wing-body configurations, where the wings and fuselage merge into a single, smooth contour. This approach avoids right angles and discontinuities in the air-frame, significantly reducing radar reflectivity. Additionally, internal payload bays are used to house weapons or sensors that would otherwise protrude from the surface and create radar hot-spots. Even small UAVs or drones used in hybrid and asymmetric conflicts are beginning to adopt simplified versions of these principles—covering payloads within the frame, using rounded air-frames, or flying at angles that reduce radar exposure.

Beyond shaping, the material composition of a UAV plays a crucial role in its radar signature. Traditional metallic air-frames, such as those made of aluminium or titanium, are highly reflective and thus easily detected by radar. In contrast, non-metallic composite materials such as carbon fibre-reinforced polymers, fibreglass, and radar-transparent polymers absorb or transmit radar energy, thereby reducing the amount of reflected signal. Some composites can be engineered with specific dielectric properties that make them nearly invisible to certain radar frequencies. Additionally, many stealth UAVs are coated with radar-absorbent materials (RAM), which contain ferromagnetic or dielectric compounds that convert incoming radar energy into heat or disperse it through internal scattering. These coatings are often tuned to absorb specific radar bands used by common military search and tracking radars. RAM can take the form of specialized paint, foam layers, or embedded particles in a resin matrix. The effectiveness of RAM increases when it is applied in combination with geometric shaping, allowing the UAV to remain undetected across a wider range of angles and frequencies.

Surface smoothness and continuity also affect radar wave behaviour. Any protrusion, seam, or discontinuity in the UAV's skin can act as a corner reflector, returning a strong radar signal. For this reason, stealth UAVs are often constructed with seamless or flush-mounted surfaces, concealed fasteners, and embedded antennas. Hatches, vents, and access panels are carefully integrated to avoid sharp transitions, and the leading and trailing edges of wings are often serrated or swept to scatter radar waves. Special attention is given to the alignment of surface features to avoid resonance at specific radar wavelengths, which could otherwise amplify the UAV's signature.

As UAV stealth technology evolves, so too do radar detection methods. Adversaries have adapted by deploying more advanced and diverse radar systems designed to detect even low-observable objects.

Synthetic Aperture Radar (SAR), for instance, uses the motion of the radar platform—such as a satellite or aircraft—to create high-resolution images of ground targets. SAR is capable of detecting stationary or slowly moving UAVs that traditional radar might miss, and it can operate in all weather conditions, including through clouds or at night. This makes SAR a powerful tool for detecting UAVs that rely solely on optical or infrared stealth.

Multistatic radar systems represent another leap in detection capability. Unlike mono-static radar, where the transmitter and receiver are co-located, multistatic radar uses multiple receivers placed at different locations. This spatial diversity allows for the detection of objects that might be invisible from one angle but visible from another. Multistatic systems are particularly effective against stealth aircraft and UAVs because they can capture scattered radar returns that would otherwise be lost. These systems can also be networked and integrated with AI to improve detection probability and tracking continuity in complex airspace environments.

AI-enhanced radar tracking is rapidly changing the balance between stealth and detection. Machine learning algorithms can be trained to recognise subtle anomalies in radar returns, enabling radar operators to distinguish between natural clutter and low-signature UAVs. These systems can integrate inputs from radar, electro-optical sensors, infrared cameras, and acoustic detectors to create a fused detection picture. UAVs attempting to evade radar might succeed against a single radar set, but in an integrated defence system guided by AI, the probability of detection increases significantly. This has led UAV designers to pursue not only passive stealth measures but also dynamic countermeasures—such as real-time RCS management, route adjustment, and electronic jamming to delay or deny detection.

Some UAVs are now being equipped with radar-transparent covers or domes that protect internal components without contributing to radar signature. These domes are often made from specialised polymers or composites with low permittivity, allowing radar waves to pass through rather than reflect off the surface. This is especially useful in nose cones or sensor housings, where the radar needs to function outward without increasing the aircraft's signature. Furthermore, electromagnetic interference (EMI) shielding is used within the UAV's electronics bay to prevent emissions from internal components (e.g., processors, radios, power systems) from leaking out and creating a detectable electronic footprint.

The rise of loitering munitions and swarm drones has also prompted the miniaturisation and simplification of RCS-reduction strategies. While these smaller UAVs cannot afford the complexity of full-spectrum stealth, they do employ low-RCS shapes, non-metallic construction, and sometimes even RAM coatings in critical areas. Their very size becomes a stealth asset, as many fall below the minimum detection threshold of traditional radar systems. However, swarms are particularly vulnerable to wide-area radar systems that use machine learning to identify and track multiple low-RCS targets simultaneously. This has driven research into group stealth dynamics, where drone formations deliberately alter shape, speed, and altitude to mask the number and nature of UAVs in flight.

In conclusion, the science of radar reflectivity and radar cross-section management is central to the ongoing development of stealth UAVs. By understanding how radar waves interact with surfaces, materials, and shapes, engineers can design UAVs that are significantly harder to detect and track. Through a combination of structural shaping, advanced composites, RAM, and emission control, UAVs can operate in hostile airspace with reduced risk of interception. However, the continual evolution of radar technology—especially AI-enhanced and multi-angled detection systems—ensures that stealth remains a moving target, requiring constant innovation. As the interplay between detection and evasion intensifies, the science of RCS reduction will remain a cornerstone of both UAV design and modern air-power strategy.

9.1.6 UAV Stealth Technology Convergence

The convergence of stealth strategies across these multiple spectrum is giving rise to a new generation of UAVs that are difficult to detect and even harder to counter. A growing area of research is the integration of multi-spectral stealth materials—engineered to simultaneously minimise detectability across radar, IR, visual, and acoustic domains. These materials combine RAM properties, thermal regulation, visual camouflage, and noise dampening into single-layer or multi-layer skins. For example, nano-composite coatings are being developed that use graphene, carbon nanotubes, and metal oxides to selectively absorb and scatter electromagnetic energy while also providing thermal conductivity and UV protection. These coatings not only improve stealth but also enhance durability and weather resistance.

Another promising innovation is the use of meta-materials—engineered structures that interact with electromagnetic waves in unconventional ways. Unlike traditional materials, meta-materials can be designed to bend, absorb, or redirect radar waves regardless of the material's base composition. UAVs equipped with meta-material coatings or structural panels may be able to achieve unprecedented levels of radar stealth, especially when combined with shape-based RCS reduction. Some theoretical meta-material configurations have even demonstrated potential for cloaking effects at certain frequencies, although practical implementation remains at an early stage.

In summary, the integration of stealth technologies into UAVs has become a multidimensional and highly technical endeavour, aimed at minimising detectability across the full electromagnetic and acoustic spectrum. From the use of radar-absorbent materials and deflective shaping to infrared suppression, visual camouflage, and noise reduction, modern UAV modified hobby design is defined as much by its ability to evade detection as by its payload or endurance. As materials science, computational design, and adaptive technologies continue to evolve, the future of UAV stealth lies in systems that are not only structurally optimised but also dynamically responsive to the changing environments and threats they encounter. In this way, stealth has become not just a defensive measure but a proactive enabler of operational dominance in the increasingly contested domains of modern warfare.

CHAPTER 13

Evolution of Drones in the Ukraine War (2014-2025)

10.1 The History and Development of Drones

The evolution of hobby drones in the Ukrainian war represents one of the most significant examples of rapid civilian technological adaptation in the modern battlefield. Beginning with the informal use of consumer quadcopters in 2014, Ukraine's experience demonstrates how inexpensive, widely available technologies can evolve from civilian novelty into indispensable military assets. Over eleven years, the integration of hobbyist, volunteer, and industrial innovation transformed drones from simple observation tools into precision-guided, autonomous, and semi-autonomous systems forming the backbone of tactical intelligence, surveillance, reconnaissance (ISR), and strike capability.

The earliest phase of drone use in Ukraine, from 2014 to 2015

The earliest phase of drone use in Ukraine, from 2014 to 2015, coincided with the initial stages of the Donbas conflict and the annexation of Crimea. Ukraine lacked a substantial military drone capability at that time, relying primarily on outdated reconnaissance systems inherited from the Soviet era. In contrast, Russia already maintained a modest fleet of military-grade UAVs. Against this technological imbalance, Ukrainian volunteers and civilian engineers turned to commercial hobby drones to fill the gap. Platforms such as the DJI Phantom 1 and 2, released in 2013–2014, became critical tools for basic reconnaissance. These drones, typically equipped with 1080p cameras and GPS-assisted flight, offered stability and intuitive controls, allowing even inexperienced operators to capture battlefield imagery. Volunteer groups, notably Aerorozvidka, played a central role during this phase. Initially formed by enthusiasts and IT professionals, Aerorozvidka pioneered the adaptation of consumer drones to wartime functions. They replaced the standard GoPro cameras with lightweight digital video transmitters, integrated 5.8 GHz analogue FPV links, and extended flight times through custom Li-Po battery packs.

These early adaptations marked the beginning of an iterative process driven by necessity. The original DJI Phantoms offered a flight duration of approximately 10–15 minutes, with limited control range. Volunteers quickly began modifying antennas, power systems, and video equipment to extend range and reliability. Using high-gain 2.4 GHz patch antennas and 5.8 GHz directional video receivers, they achieved practical operating distances of up to two kilometres, providing valuable real-time reconnaissance for front-line artillery and infantry units. The simplicity of the Phantom's flight controller—essentially an early variant of the NAZA system—proved both a limitation and a strength. Its stability in hover and automatic return-to-home functions allowed operators to focus on visual intelligence rather than complex piloting tasks.

2016 Ukrainian volunteer

By 2016, the Ukrainian volunteer movement had begun to formalise. Civilian workshops and maker spaces transitioned from field improvisation to small-scale production. Hobby drones diversified into two dominant forms: multi-rotor and fixed-wing. The multi-rotor class, represented by quad-copters and hex-copters, provided agility, hovering capability, and vertical takeoff and landing—ideal for close-range reconnaissance and observation over urban and forested terrain. Fixed-wing hobby aircraft, often based on foam models such as the Skywalker or the MyFlyDream series, delivered greater endurance and range, capable of operating beyond visual line of sight (BVLOS) for up to 60 minutes or more. These aircraft typically used open-source flight controllers such as ArduPilot and INAV, running on affordable micro-controllers like the STM32 series.

The adaptation of open-source autopilot systems represented a pivotal shift in the technological development of Ukrainian drones. Unlike proprietary DJI systems, open-source platforms allowed full customisation. Ukrainian engineers modified firmware to optimise way-point navigation, loiter patterns, and autonomous return functions. Ground control stations based on Mission Planner or QGround-Control software enabled real-time telemetry and mission planning via 433 MHz or 900 MHz radio links. This level of autonomy made fixed-wing drones invaluable for mapping enemy positions and coordinating artillery strikes. The use of Pixhawk-based controllers, combined with GPS and magnetometer modules, became the standard for semi-autonomous reconnaissance flights.

Propulsion evolution

In parallel, battery and propulsion technology also evolved. The period between 2016 and 2020 saw substantial improvements in Li-Po cell chemistry and discharge rates. Early 3S (11.1 V) configurations gave way to 4S and 6S systems, providing higher voltage for brush-less motors, improved efficiency, and longer flight duration. Motors such as the 2207 and 2306 series—originally designed for FPV racing—offered exceptional thrust-to-weight ratios and durability. Electronic speed controllers (ESCs) improved with higher amperage ratings and active braking firmware, enabling faster response times. Together, these improvements turned the humble hobby drone into a stable, reliable reconnaissance tool capable of surviving the rigours of the front line.

Second phase of development, between 2017 and 2021

The second phase of development, between 2017 and 2021, also coincided with the growth of the global FPV (First-Person View) racing community. FPV technology, originally designed for sport and recreation, would later form the backbone of Ukraine's front-line drone capabilities. Racing drones used low-latency analogue video transmission, high-thrust motors, and lightweight carbon-fibre frames. Ukrainian volunteers recognised that these characteristics made FPV platforms ideal for close-range tactical reconnaissance and precision manual flight. By 2020, workshops across Ukraine were producing improvised FPV systems capable of carrying small explosive payloads.

2020 and 2021, fixed-wing development

In 2020 and 2021, fixed-wing platforms matured further through the integration of digital video systems and enhanced telemetry. The transition from analogue 5.8 GHz video to digital systems, such as DJI FPV and later Walksnail or HDZero, provided clearer imagery and reduced interference. Ground stations improved through diversity receivers and directional antennas, increasing range and situational awareness. Flight planning became increasingly sophisticated, with Ukrainian engineers developing pre-programmed waypoint missions that allowed aircraft to operate autonomously for up to 80 kilometres—depending on weather and signal interference.

Early 2022

By early 2022, with Russia's full-scale invasion, Ukraine's informal drone network transformed into a critical part of the nation's defence architecture. Volunteer workshops became semi-industrial production lines. FPV drones, particularly those built around 5-inch racing frames and running Beta-flight firmware, emerged as one of the war's defining innovations. Operators adapted existing 5-inch FPV drones to serve as precision loitering munitions. These drones were manually guided to their targets via FPV goggles, transmitting low-latency analogue video feeds. Powered by 6S 1500 mAh Li-Po batteries, they offered approximately five to seven minutes of flight time—sufficient for tactical strikes within a two-to-three-kilometer radius.

Accelerated adaptation

Technological adaptation accelerated rapidly through 2022 and 2023. Ukrainian units standardised production of FPV drones, integrating lightweight explosives—typically repurposed RPG warheads, VOG grenades, or 3D-printed casings containing TNT equivalents. The racing components industry, already globalised and accessible, allowed Ukraine to source motors, frames, and flight controllers despite wartime restrictions. Many of these drones were hand-assembled using off-the-shelf parts: carbon-fibre arms, 30×30 mm stack-mounted flight controllers, and micro FPV cameras. The affordability and modularity of these systems allowed mass deployment and easy replacement.

Simultaneously, consumer drones like the DJI Mavic 2, 3, and Mini series found new life on the battlefield. The Mavic's high-resolution cameras, foldable design, and GPS stabilisation made them invaluable for reconnaissance, artillery spotting, and target verification. Despite their civilian limitations—particularly geo-fencing restrictions and radio vulnerabilities—Ukrainian operators developed workarounds. Software modifications disabled flight restrictions, while signal boosters and aftermarket controllers extended range. The Mavic 3's 4K camera allowed for highly detailed imagery, supporting both tactical and strategic intelligence collection.

Fixed-wing drones also evolved in complexity and endurance. Hobby-grade foam aircraft were reinforced with carbon-fibre spars, equipped with high-efficiency brush-less motors, and powered by larger 6S or even 8S Li-Po battery configurations. These fixed-wing systems could remain aloft for up to 90 minutes, covering distances exceeding 60 kilometres. They served as mapping platforms, signal relays, and reconnaissance tools capable of penetrating deep into contested airspace. Flight controllers running ArduPilot or PX4 firmware offered full mission autonomy, including way-point navigation, altitude hold, and automatic landing sequences.

Late 2023

By late 2023, Ukraine's drone innovation reached industrial scale. FPV and fixed-wing systems were being manufactured in decentralised workshops nationwide. Some specialised in frame construction using 3D printing or CNC machining, while others focused on electronic integration and flight tuning. The modular nature of hobby drone components allowed rapid prototyping and field repair. The drone ecosystem also diversified: 3-inch "micro" FPV units were used for urban reconnaissance, 5-inch platforms for medium-range attack, and 7-inch or 10-inch long-range builds for extended flight and heavier payloads.

The conflict also spurred the development of hybrid platforms combining multi-rotor and fixed-wing features. Vertical Takeoff and Landing (VTOL) designs emerged from the maker community, blending quad-copter lift systems with efficient fixed-wing cruise modes. Although technically demanding, these designs offered flexibility in rugged environments with limited launch space. Early Ukrainian prototypes employed dual propulsion modes—vertical lift motors powered by 4S Li-Po batteries and a rear pusher motor driven by a 6S system for forward flight. These systems achieved endurance of up to two hours while maintaining compact transportability.

Between 2023 and 2025

Between 2023 and 2025, the technological sophistication of Ukraine's hobby-drone-based systems continued to escalate. Digital FPV systems became increasingly common, using 2.4 GHz or 5.8 GHz digital transceivers for encrypted video and telemetry. Low-latency HD video feeds enhanced operator accuracy during FPV strike missions. The rise of AI-assisted targeting—supported by onboard processors such as Raspberry Pi and Nvidia Jetson Nano boards—allowed semi-autonomous identification of vehicles and personnel. These systems processed live camera feeds, comparing patterns to stored image databases for rapid target recognition.

Meanwhile, electronic warfare (EW) became a defining factor in the survival of drone operations. Both sides invested heavily in radio jamming and spoofing technologies. Ukrainian drone designers responded by integrating frequency-hopping spread-spectrum (FHSS) systems, directional antennas, and redundant control links. Some drones included inertial navigation systems (INS) as backups for GPS-denied environments. Others incorporated Starlink satellite links for long-range command relay, though this was limited by bandwidth and latency.

By 2024

By 2024, Ukraine's drone industry had matured into a layered ecosystem. Civilian suppliers produced frames and components, while military-affiliated workshops focused on assembly and deployment. The government formalised many volunteer initiatives, providing funding, standardisation, and logistics. FPV drones became central to Ukraine's asymmetric warfare doctrine. A single operator could train in under a week, assemble a drone in a day, and deliver precision strikes at a fraction of the cost of traditional munitions.

In 2025

In 2025, fixed-wing and hybrid drones were increasingly networked into integrated systems. Swarm experiments began, where multiple drones operated in coordination using simple mesh communication protocols. Artificial intelligence facilitated target allocation, formation flight, and threat avoidance. Energy density improvements in Li-Po and Li-Ion hybrid batteries extended operational time, while composite materials enhanced durability without significant weight penalty.

The transition from hobby drones to tactical assets reshaped warfare itself. The Ukrainian conflict revealed that inexpensive, adaptable, and locally produced drones could offset conventional disadvantages in manpower and equipment. Each phase of technological evolution reflected Ukraine's adaptive capacity: the early improvisation of 2014, the open-source integration of 2016–2020, the industrial acceleration of 2022, and the systematisation of 2023–2025.

Multi-rotor platforms offered flexibility, ease of use, and real-time responsiveness, making them indispensable for close-range ISR and precision strikes. Fixed-wing systems provided endurance, autonomy, and operational depth, enabling persistent surveillance and intelligence gathering. The coexistence of these two classes, coupled with the ingenuity of volunteer engineers, defined the unique character of Ukraine's drone warfare.

As of late 2025, the Ukrainian model has inspired global military transformation. Nations across Europe and Asia have begun integrating FPV-style systems into formal military doctrine, recognising their cost-effectiveness and tactical agility. What began as a volunteer effort with hobby drones has evolved into a paradigm shift in modern conflict—where accessibility, adaptability, and innovation outweigh industrial scale alone.

The Ukrainian experience underscores that the boundary between hobbyist and military technology has effectively dissolved. Civilian hardware, once designed for sport and leisure, now occupies a central role in warfare. The fusion of open-source software, hobby electronics, and field innovation represents not only a technical achievement but a social and strategic transformation—a democratisation of aerial warfare, born from necessity and sustained by resilience.

10.2 (3D-printed) drone parts in the Ukraine

From filament to fleets: the development of 3D-printed drone parts in the Ukrainian war (date-ordered narrative)

The story of 3D-printed drone parts in Ukraine begins as a modest, improvised response to necessity and supply constraints and evolves, within a few years, into a layered national capability spanning hobbyist garages, volunteer maker collectives, and semi-industrial production lines. That arc — from tinkering to scaled production — mirrors the broader adaptation of civilian manufacturing technologies to wartime conditions, where speed of iteration and distributed capacity mattered as much as raw engineering sophistication.

2014–2016: the first experiments and maker responses

In the immediate aftermath of the 2014 Donbas fighting, Ukraine's technological volunteers were focused first on information, medical supplies and low-tech improvisation. 3D printing at that stage was largely embedded in maker communities and universities rather than front-line workshops. Early adopters experimented with printed camera mounts, simple brackets, small enclosures and prototype jigs to support consumer quad-copters pressed into service for observation and documentation. The printers in use were predominantly desktop Fused Deposition Modelling (FDM) machines: RepRap-style kits, early Prusa i3 iterations and consumer models from nascent Chinese manufacturers. These machines were valued for their simplicity, low operating cost and the ability to produce bespoke plastic components on demand.

At this phase, printed parts were conservative in scope: camera housings, vibration dampers (non-structural), adaptor plates for third-party cameras, and lightweight cable management clips. The priority was rapid fit-and-replace capability rather than structural innovation. Materials were typically PLA (polylactic acid) for prototypes and PETG for parts requiring marginally better temperature and impact resistance. The community emphasis was on iterative design — print, test, revise — an approach that would remain defining even as printing technologies matured.

2016–2019: makerspaces, diversity of parts, and gradual standardisation

Between roughly 2016 and 2019, Ukraine's makerspaces and volunteer workshops proliferated and began to specialise. The maker movement, including university FabLabs and independent groups, introduced more capable desktop printers (notably Prusa i3 MK2/MK3 series and Ender/Creality series machines) and broadened material use to include ABS and tougher PETG blends. With improved printers and filament choices, volunteers expanded the catalogue of printed drone parts.

During this period, printed components included: protective housings and fairings for consumer air-frames, lightweight non-structural frame reinforcements (clips and braces), landing gear skids, battery retainers designed for specific packs, camera gimbals and mount adaptors, antenna holders, and custom mounts for telemetry radios and small sensors. Many of these parts were designed to be modular and easily swapped in the field, addressing logistical problems such as damaged feet, broken mounting tabs, or unavailable factory spares. Desktop SLA (resin) printers began to appear in specialist workshops for producing fine-detail components — small optical housings, lens mounts, and precisely fitting connector plugs — though their use remained limited because SLA resin parts generally lack the toughness required for high-stress field applications.

Two important dynamics emerged in this phase. First, open-source design sharing accelerated: file repositories, community forums, and Telegram channels enabled designs to spread rapidly between towns and regions. Second, volunteer groups began to coordinate production runs for specific parts — not yet mass manufacturing, but organised batches to meet urgent needs. The transition from ad-hoc prints to coordinated production rehearsed the networked production models that would scale in 2022.

2020–2022: improved materials, stronger desktop machines, and prototype composites

The years before the full-scale invasion saw several technological inflections relevant to printed drone parts. Desktop printers continued to improve in reliability and print quality; models like the Prusa i3 MK3 gained traction for their robustness in harsh conditions. More importantly, new filament formulations — carbon-fibre filled nylons, high-temperature PETG blends, and engineering-grade polyamides — became accessible to advanced hobbyists and small workshops. These materials opened practical, non-actionable avenues for producing tougher housings and serviceable replacement components that better tolerated heat, wear and mechanical stress.

Meanwhile, a small but critical number of Ukrainian groups experimented with higher-end additive technologies. On the one hand, resin stereolithography (SLA) printers like those from Formlabs were used for precision fixtures and small connectors. On the other hand, composite filament printers and desktop continuous-fibre solutions (the earliest rows of reinforced filament systems) were trialled for parts that needed superior stiffness. These were still niche, but they signalled an important shift: printing was no longer only for cosmetic or lightweight applications; with better materials and post-processing, 3D-printed parts could play more durable, serviceable roles in fielded drones.

February 2022 and the pivot to urgency: scaling the network

Russia's full-scale invasion on 24 February 2022 precipitated a rapid reallocation of technological capacity across Ukraine. The volunteer 3D printing community transformed from a dispersed network of hobbyists and med-tech producers into an explicitly defence-oriented production ecology. Groups such as "3D Print Army," "DrukArmy" and other volunteer collectives coordinated thousands of printers — both local and donated — to produce mission-support items. At the same time, international actors in the maker ecosystem, notably Prusa Research, reported supplying printers and material assistance to Ukraine, reinforcing the domestic base. This distributed production model allowed immediate, localised manufacturing of components that met urgent operational needs.

The character of printed drone parts changed markedly in 2022. Speed of iteration became paramount: volunteers printed replacement brackets, transmitter mounts, custom gimbal adaptors, prop guards, and repair spares — items that kept fleets of consumer and racing drones operational without waiting for supply chains to reconstitute. The most common machines remained desktop FDM units (Prusa MK3, Creality Ender 3), prized for their ubiquity and the enormous community knowledge base surrounding their operation. Their advantages were simple: low purchase cost, broad spare-part availability, and a huge catalogue of validated print profiles that reduced the learning curve for new operators.

At the same time, a parallel layer of capability surfaced: small commercial workshops and startups began to operate clusters of more capable machines. These included larger format FDM printers, resin SLA units, and, in some regional hubs, industrial composite printers from manufacturers such as Markforged. The Markforged line offered, in particular, the capability to print continuous-fibre-reinforced parts (nylon matrix with carbon or glass fibre reinforcement) that combined higher stiffness and toughness than ordinary thermoplastics — properties useful when stronger, more durable mounts or structural sub-components were required. Reporting and industry commentary showed how firms and better-resourced workshops used such machines to produce robust, repeatable parts at a tempo beyond what hobbyists could sustain.

2022–2023: production breadth and the catalogue of printed drone parts

Through 2022 and into 2023, the inventory of printable drone components expanded dramatically. Volunteer and semi-industrial producers supplied thousands of part types to drone teams and mechanised units. The most commonly printed items included:

- Replacement mounting plates and frame brackets — pieces allowing fusion between proprietary consumer components and locally available hardware.
- Camera and gimbal housings — custom housings to fit different small cameras and to add vibration management (non-structural damping).
- Antenna and radio mounts — holders and directional antenna collars that enabled field attachment of

extended-range telemetry gear.

- Landing gear and skid replacements — ruggedised skids and low-cost sacrificial feet for multi-rotors operating in rough terrain.
- Payload enclosures and sensor pods — weather-resistant housings to carry small non-weaponized sensors, lights, or environmental monitors.
- Repair jigs and testing fixtures — ground-support equipment used to test motor rotation, ESC function and to hold air-frames during maintenance.
- Cable clips, battery retainers and quick-release hardware — logistical small parts that materially improve operational turnaround time.

Critically, most of these parts were deliberately designed to be non-specialist and easily replaceable. Volunteer designers emphasised modularity so parts could be swapped quickly in the field. Community repositories — private and public — circulated STL files and CAD models, while Telegram and Discord channels arranged local production runs to fulfil unit requests.

Not all printing was performed on low-cost desktop machines. As demand grew and requirements hardened, some workshops transitioned to industrial printers for specific use cases. Small contract manufacturers and startups used technologies such as selective laser sintering (SLS) and industrial FDM with higher performance polymers to produce components that needed better heat resistance, wear properties, or a smoother surface finish. These machines, often supplied and operated by firms inside and outside Ukraine, enabled larger batch runs of consistent parts suitable for extended operational use. Reporting from industry observers documented facilities where rows of printers churned out chassis clips, brackets and housings at scale — evidence of an emergent semi-industrial supply base.

2023–2024: composite printing, integration and quality focus

By mid-2023 a discernible pattern had emerged: desktop printing supplied vast quantities of soft-use parts and emergency spares, while higher-grade printers handled components requiring durability and repeatability. Composite printers — either continuous-fibre reinforcement systems or filament extruders using carbon-fibre filled nylon — became more common in specialised shops. The outputs from these machines were not general structural air-frame replacements, but rather targeted components that benefited from better stiffness or thermal performance: secure mounts for heavier cameras, hardened antenna collars, and reinforced motor mounts used in larger custom multi-rotors. These parts were typically post-processed and quality-checked before fielding. Industry accounts describe Markforged and similar systems enabling functional composite parts to be delivered faster than conventional manufacturing routes, particularly when supply chains were disrupted.

Another salient development in 2023 was the formalisation of quality practices. As drones became mission-critical, workshops adopted inspection protocols, consistent material sourcing, and version-controlled designs. Print farms producing hundreds or thousands of parts implemented simple QA steps: visual inspection, fit checks against reference parts, and basic environmental testing (exposure to heat or UV in simulated conditions). These practices drove up the reliability of printed parts and expanded the range of acceptable field applications.

2024–2025: scale, specialisation and industrial partnerships

In the period 2024–2025 the Ukrainian 3D printing ecosystem matured into a multi-tier supply chain. Dozens of small firms specialised in production of drone components, offering catalogues of prints optimised for specific local platforms. Some companies, profiled by industry outlets, established dedicated production lines with rows of matched printers and standardised post-processing stations; others focused on R&D, iterating stronger composite prints and more thermally stable housings. The narrative shifted from emergency improvisation to supply-chain engineering: design for manufacturability, batch control, and life-cycle replacement planning.

Notably, several private enterprises and reporting outlets highlighted the use of 3D printing to accelerate product development cycles for new drone models. Rapid prototyping allowed designers to validate new mounts, sensor enclosures and chassis geometries within days rather than weeks. Where a prototype graduate to production, larger printers or moulding methods were sometimes introduced — but the speed advantage of additive manufacture was the decisive factor for iterative improvement under wartime pressures. Industry reporting, and interviews with company founders, described facilities where printed clips, braces and seat-of-the-pants components moved from prototype to standardised production in short order.

Printers and processes in use: a typology (non-actionable summary)

Across this timeline, three broad printer categories became central to Ukraine's printed-parts ecosystem:

Desktop FDM (open-frame and closed-frame machines): examples include Prusa i3 MK2/3 series, Creality Ender series, and similar consumer machines. These were the backbone of distributed production — ubiquitous, low-cost, and easy to maintain. They printed PLA, PETG, ABS and carbon-filled filaments in small batches.

Resin SLA/DLP printers: Formlabs and hobby resin machines were used for fine-detail components, lens housings, and small connectors that require high dimensional fidelity. Their parts generally served non-high-stress roles unless post-processed in robust resins.

Industrial and composite printers: Markforged, larger industrial FDMs, and SLS systems (e.g., EOS class equipment) were used in dedicated workshops for tougher parts. These enabled nylon-based prints, continuous-fibre reinforcement and consistent batch production for components requiring better mechanical performance.

Across these categories, the common thread was pragmatic material choice. PLA and PETG dominated for rapid parts; ABS or ASA were used when UV resistance was needed; Nylon and carbon-filled nylons were reserved for reinforced pieces produced on higher-capability machines.

Ethics, safety and the boundary of dual use

Throughout this evolution, Ukrainian makers and firms grappled with ethical and legal frames. Much of the volunteer printing focused on lifesaving or mission-support roles: medical fixtures, personal protective equipment, vehicle adaptations and non-lethal drone support parts that kept ISR platforms airborne. Public commentary and media reporting also noted the troubling dual-use potential of 3D printing: the same capability that yields quick replacement parts can be repurposed for weapon systems or munitions production. This reality pushed some suppliers and international donors to set explicit boundaries on what designs and equipment they would support, emphasising humanitarian and defensive applications.

Lessons learned and the lingering legacy (2025 outlook)

By 2025 the 3D-printing story in Ukraine had settled into a multi-layered model. Distributed desktop printers provided an agile, local buffer against supply chain interruptions; small contract shops supplied higher-grade parts and repeatable batches; and a handful of more capable industrial setups offered composite and SLS parts where durability mattered. Together, these layers created resilience: front-line teams could request rapid local prints for a failed camera mount in the morning and receive a field-ready replacement by afternoon, while more durable reinforced parts moved through semi-industrial channels.

The war accelerated the normalisation of additive manufacturing in defence contexts — not because 3D printing solved every problem, but because it dramatically shortened feedback loops between design and use. Designers could see how a printed part performed in an operational environment and iterate within days. That ability to learn fast, adapt designs, and route production to wherever printers were available proved decisive in a logistics environment where traditional procurement and manufacturing channels were slow or unavailable.

Summary (non-actionable synthesis)

The development of 3D-printed drone parts in the Ukrainian war is a story of distributed ingenuity, rapid iteration and graded industrial maturation. It began with desktop printers in maker spaces and ended — at least for now — with a layered ecosystem capable of supplying thousands of non-specialist parts on demand and producing hardened components in specialised facilities. Along the way, the community refined design practices, embraced better materials, and adopted production protocols that improved reliability. That evolution demonstrates how additive manufacturing, when matched with motivated human networks and pragmatic design discipline, can materially increase resilience in contested environments — while simultaneously raising nontrivial ethical and regulatory questions about dual use and proliferation.

10.3 From Kits to Combat

2014 — The first improvisations: consumer "ready-to-fly" quad-copters

The earliest and most visible class of hobby kits pressed into wartime service in Ukraine were the consumer ready-to-fly quad-copters introduced to the market in 2012–2014. These were compact, foldable or single-piece multi-rotors sold complete with integrated flight control, camera gimbal and autopilot-like stability assists. Early models in practical use included first-generation DJI Phantoms and nascent Mavic-class designs and their many Chinese equivalents. Their appeal in 2014 was simple: they were inexpensive relative to military ISR assets, widely available, easy to fly in stabilised modes, and carried on-board cameras capable of recording and transmitting imagery useful for situational awareness. Volunteer groups and individual hobbyists — often with backgrounds in photography, IT or radio control modelling — learned to operate these kits under austere conditions, using them for observation, documentation, and basic mapping. The civic volunteer collective Aerorozvidka emerged in this period as a focal point for adapting these consumer systems to front-line reconnaissance roles, coordinating pilots and sharing early lessons about range, endurance and robustness.

2015–2016 — Modularisation consumer kits and bridging to hobby components

As the conflict persisted, the character of adaptations shifted from pure use toward modular modification. Operators began to mix and match components from hobby kits with aftermarket hobby parts to improve mission fit. The chassis and flight controllers of consumer kits were often retained for their stability and ease of operation, while external peripherals — camera options, mounts, and antenna solutions — were swapped to better suit local needs. At the same time, hobbyist multi-rotor kits built from separate frames, power systems and controllers started to appear on the front lines: these are the "frame + stack" style hobby kits beloved by RC hobbyists. Rather than a single integrated consumer unit, these kits arrived as discrete items (a carbon frame, motors, ESCs, battery, flight controller and radio receiver) that allowed field engineers to configure platforms with a mix of endurance or agility depending on the tactical requirement. The key change in 2015–2016 was therefore flexibility: units could now be field-tailored by replacing a single failed sub-assembly instead of discarding an entire consumer drone. Community knowledge-sharing made these techniques available to new operators across regions.

2016–2018 — The rise of racing FPV kits and dedicated hobby frames

Between 2016 and 2018 an important, parallel hobby ecosystem matured: the FPV racing community. Racing kits — sold as complete bundles or as modular component sets — centred on lightweight carbon frames (3"–5" propeller classes), high-KV brush-less motors, compact flight controller stacks optimised for low latency and analogue FPV camera/transmitter packages. These kits were designed for speed and manual piloting rather than stable hover photography; they offered a very different performance envelope from consumer Phantoms or Mavics. Ukrainian tinkerers recognised that the FPV kits' attributes — agility, compact size and extreme responsiveness — could be valuable in tactical contexts where manual precision or rapid ingress/egress mattered. Throughout 2018 these kits were adopted mainly for reconnaissance where speed mattered, and the FPV stack (camera, video transmitter, goggles) entered the vocabulary of Ukrainian drone teams. Public reporting later documented the role of racing-derived systems in fast, precise missions during the larger 2022 conflict phase.

2017–2019 — Fixed-wing hobby kits reappearing for endurance tasks

While the multi-rotor story dominated press narratives, a quieter but strategically important trend was the renewed use of fixed-wing hobby kits. Fixed-wing kits — foam trainers, slow-fly gliders and electric pusher designs — offered more efficient lift and thus longer endurance for a given battery mass. Hobby fixed-wing kits were commonly sold as ARF (almost-ready-to-fly) or kit-plane packages, including foam air-frames, brush-less motors, propellers and basic autopilot-compatible mounting points. Ukrainian groups employed these kits for mapping, long-range observation, and persistent area surveillance. The conversion pattern here emphasised autonomy and mission planning: hobby fixed-wing kits were combined with open autopilot stacks and ground control software to execute way-point surveys, producing orthophotos and reconnaissance tracks. The fixed-wing kits filled the endurance gap that multi-rotors could not, especially in less cluttered terrain.

2019– 2021 — Consolidation: component FC & ESC off-the-shelf expansion

As hobby-class UAS became a semi-standard tool within Ukrainian volunteer and military adjacent units, a hybrid market of component kits proliferated. These were not only "complete drones" but defined assortments targeted at specific missions: long-endurance photography kits, lightweight FPV racing kits, and "survey" fixed-wing kits. Each kit typically included a frame, motor set, ESCs, battery recommendations, a flight controller board (with varying levels of built-in capability), and either a camera or a camera mount. Flight controllers and firmware ecosystems matured in this period — hobby projects such as Betaflight, iNav, ArduPilot and others provided community-vetted options for different kits. The conversion pattern became one of firmware selection and configuration: operators chose the kit whose hardware matched the mission and then selected the appropriate controller software for stability, way-pointing, or low-latency manual control. Importantly, this period entrenched the modular mentality: mission planners could choose a kit, specify a controller stack and select peripheral radios and cameras to match constraints.

2022 onward — Rapid scaling, FPV adoption and hybridisation

The full-scale invasion that began on 24 February 2022 dramatically accelerated conversion and scaling. FPV racing kits transitioned from a marginal hobby niche into a core tactical tool; component kits that had previously been used for sport became standardised, produced in higher volumes, and distributed through volunteer networks and semi-industrial workshops. The principal classes of kits in widest use were: (1) consumer presume kits (Mavic/Phantom-style) retained for stable ISR; (2) FPV racing kits (3"–5" frames) used for rapid, precise manual missions; and (3) fixed-wing survey kits employed for longer endurance mapping and relay missions. Conversion in 2022 emphasised three high-level aims — resilience, range, and sensor fit — and the ways operators achieved those aims reflected the nature of the underlying kits. For consumer kits the focus was on telemetry integration and geo-tagging; for FPV kits the emphasis was on pilot training, control fidelity, and durable frames that could be field-repaired; for fixed-wing kits the priority was mission autonomy and payload stability. The result was an ecosystem where basic kits could be repurposed at scale for many non-lethal battlefield functions. Reporting and defence analyses highlighted FPV racing kits' outsides role in tactical operations after 2022, especially for precision manual tasks that benefited from the racing stack's low latency and agility.

2022–2023 — How hobby kits were adapted (high level)

Descriptions of adaptation in open sources focus on functional outcomes rather than technical procedures. Across kit types, adaptation typically involved choosing a baseline kit that matched the mission envelope (endurance vs speed vs stability), then integrating alternative sensors or operator interfaces and applying more robust logistics patterns to maintain fleets. For consumer kits the conversions were often pragmatic: replacing fragile cosmetic parts with print or machined replacements, augmenting data links with longer-range telemetry radios, and improving mounting points for additional sensors. For racing kits, adaptation emphasised pilot ergonomics, survivability and rapid field repair — operators selected stronger frames, redundant prop mounts, and simplified electrical stacks to enable quick replacement. Fixed-wing kits were adapted for autonomy through the use of open autopilot stacks, enabling programmed survey flights and data capture on return. Across all adaptations the recurring themes were modularity, redundancy and the ability to repair in the field — not to change the basic flight characteristics but to make the kits more suited to sustained operational use. Public reporting describes many such patterns without providing operational instructions.

2023 — Semi-industrial kits and standardised packages

By 2023 some Ukrainian manufacturers and volunteer collectives were packaging standardised kits aimed directly at front-line requirements. These kits were built from hobby components but assembled, tested and distributed with standard operating notes and spare-part bundles. The kits aimed to streamline training and maintenance: a single kit model could be taught to a small team, spare parts could be stocked in predictable quantities, and repair procedures became routine. Standardised kits included: endurance multi-rotors with recommended prop/ motor/ battery pairings for longer flight; ruggedised FPV racing kits with reinforced arms and simplified electrical stacks for quick swaps; and fixed-wing survey kits pre-configured with autopilot stacks for mapping missions. The effect of these developments was to shorten the supply and training cycle and to make hobby kits more reliable in combat conditions. News accounts and industry reports from this period document the rise of such semi-industrialised kit suppliers and their growing production capacities.

2023–2024 — Evolution of supporting kits and logistical bundles

As kits moved from improvised assemblies toward standardised field kits, another class of supporting kits emerged: repair and logistics bundles. These are not flight platforms but inventories of spares and replacement sub-assemblies packaged to allow field teams to maintain dozens of kits without return to base. Typical inclusions in public descriptions are standardised prop packs, motor sets, spare frames or arms, tool-kits for non-specialist maintenance, and pre-flashed controller boards (the latter as a provisioning convenience). The logistics bundles reflect a maturing understanding that hobby kits are best used when they are easy to sustain: predictable repair time, available spares and trained crew. This logistics thinking turned hobby kits into a replicable capability rather than a collection of one-off hacks.

2024–2025 — Hybrid kits, new form factors and supply-chain scaling

Toward late 2024 and into 2025 the hobby kit landscape continued to diversify. Hybrid VTOL kits (combining vertical lift with winged cruise) matured in maker workshops and small firms, offering transportable endurance for environments with constrained launch space. Meanwhile, specialised chassis kits for maritime or loitering tasks (non-weaponized roles such as mine detection or signal relay) became part of the catalogue. Large-scale manufacturing capacity estimates and policy statements in 2024–2025 indicated that Ukrainian industry could produce high volumes of FPV-class kits and competing forms — a reflection of both domestic supplier maturation and increasingly formalised procurement. Reporting in 2025 documents growing domestic output and the emergence of firms that package hobby components into mission-ready kits at scale. These trends show how the original hobby kit ethos — modular, accessible and community-driven — evolved into a sustained industrial capacity.

Technical themes across kit types (non-actionable synthesis)

Across the entire period a few technical themes recur in how hobby kits were chosen and adapted. First, modularity: component-style kits (frames + stack + payload) allowed field engineers to replace only the failed sub-assembly, significantly lowering life-cycle cost and downtime. Second, standardisation: as kits were produced in larger numbers, standardised sub-assemblies and spare packs reduced training time and logistical complexity. Third, resilience: conversions emphasised redundancy and ease of repair (stronger arms, simplified wiring harnesses, use of printed replacement parts) rather than changes in the basic aerodynamics or control logic of the kits. Fourth, fit-for-purpose selection: operators chose kits based on mission needs — endurance for mapping, agility for close reconnaissance, and stability for high-quality imagery — and performed non-intrusive adaptations to match those needs. These patterns demonstrate that the "conversion" of hobby kits was primarily an exercise in systems selection, modular upgrades, and logistics engineering rather than wholesale redesign.

Ethical, legal and operational caveats

Once again, it is important to underline that converting hobby kits for use in conflict raises complex ethical and legal questions. Many of the public accounts stress that volunteer groups focused on reconnaissance, rescue support and force protection — civilian and humanitarian roles. At the same time, the dual-use character of hobby kits meant that conversion potential always carried the risk of weaponisation or escalation; this reality prompted some vendors and international supporters to draw explicit lines around what equipment and designs they would supply. Reporting and commentary on the conflict routinely highlight these dilemmas and the need for governance frameworks around dual-use technologies.

Conclusion: hobby kits as scalable, modular capability

From 2014's consumer Phantoms to the mass proliferation of modular FPV and fixed-wing kits by 2025, the Ukrainian experience maps a clear arc: hobby kits started as accessible consumer goods and evolved — through community innovation, standardisation and semi-industrial production — into a scalable and sustainable component of national resilience. The "conversions" were less about radical redesign than about selection, integration and logistics: choosing the right kit for the job, equipping it with appropriate sensors and radios, and ensuring predictable maintenance through spare-kit bundles. In that sense the story of hobby kits in Ukraine is a story about modular systems engineering and distributed manufacturing under pressure — an example of how hobbyist technologies can be repurposed ethically for defensive, humanitarian, and intelligence purposes when governed and deployed with care.

10.4 Ukraine Hobby Drones in Action.

Scenario - 1 Offensive Tactics (Assaulting Fortified Position)

From the 2014 invasion to the Trenches in 2022 - UAV's deploying grenades

From the opening months of Russia's invasion in 2014 through to the full-scale invasion in 2022, Ukrainian units and volunteer networks progressively adapted commercial "hobby" drones into a trench-level reconnaissance-strike system capable of operating from trench to trench with minimal logistics and at very low cost. Small quad-copters—particularly widely available DJI-type platforms—were first fielded in large numbers for immediate intelligence, surveillance, and reconnaissance (ISR) tasks: scanning tree lines and roof-lines, detecting vehicle movement, confirming enemy firing points, and, critically, correcting mortar and artillery fire in near real time, often within a single engagement cycle. As the front-line stabilised and fortifications expanded, these same quad-copters were increasingly modified for direct attack, using simple release mechanisms to drop small munitions—typically hand grenades or improvised explosive charges—onto troops sheltering in trenches, dugouts, and behind cover where flat-trajectory weapons were less effective.

This "vertical enfilade" effect exploited limited overhead protection and the drone's ability to hover, observe, and re-attack within minutes, generating persistent psychological pressure that forced dispersal, restricted movement, and complicated routine activities such as resupply and casualty evacuation. By late 2022, grenade-dropping quad-copters had become a normal feature of close combat, enabled by ad-hoc field engineering—including 3D-printed mounts and basic servo releases—that allowed rapid fielding, replacement, and tactical iteration at scale. Alongside these strike-adapted quad-copters, Ukraine relied heavily on fixed-wing and hybrid UAVs—a mix of volunteer-supplied and domestically produced systems—for longer-endurance reconnaissance, mapping defensive belts, cueing fires, tracking logistics routes, and providing broader situational awareness beyond the limited range and endurance of small quad-copters. Doctrinally, this evolution is best understood as the emergence of a mass, low-cost unmanned aerial systems layer that fused reconnaissance, target acquisition, and short-cycle strike at the tactical edge, reducing reliance on scarce high-end platforms and enabling small units to generate disproportionate effects through rapid sensing, rapid decision-making, and repeated precision from above. Ref: Appendix (B) example, 2

From a technical perspective, a simple quad-copter equipped with more efficient propellers and a high-quality Li-Po or Li-ion battery optimised for endurance can generate sufficient lift to carry a small payload of roughly 600 grams over a moderate distance, making short-range, repeatable missions theoretically feasible while keeping operators physically removed from immediate danger. In the early phases of the war, particularly through 2022 and into 2023, this approach became increasingly prevalent because it combined low cost, precision, and rapid redeployment, allowing small teams to exert disproportionate battlefield effects. As the conflict evolved, however, the growing use of electronic warfare significantly disrupted control links and navigation, forcing adaptations in how drones communicated and operated. Rather than eliminating this method altogether, these countermeasures accelerated innovation and diversification, with alternative control techniques emerging alongside a marked increase in one-way or expendable drone systems. As a result, the underlying logic of using inexpensive unmanned platforms to offset more expensive and complex systems not only persisted, but expanded, shaping a new and enduring pattern of drone employment on the modern battlefield.

Scenario - 2 Offensive Tactics (Assaulting Amour)

From the Trenches in 2022 - UAV's deploying RPG (Rocket Propelled Grenade)

The use of quad-copter drones with a high lift profile to deliver rocket-propelled munitions against hardened targets has emerged as a notable feature of modern irregular warfare, reflecting the convergence of inexpensive commercial technology with long-established battlefield weapons. Rather than replacing traditional anti-armour systems, these drones have been used as improvised delivery platforms, enabling small units to exert disproportionate pressure against larger and better-protected targets such as armoured personnel carriers (APCs) and fortified bunkers. Their importance lies less in raw destructive power and more in their psychological effect, adaptability, and ability to reshape how space and threat are perceived on the battlefield. At the same time, this method of attack carries significant limitations that constrain its effectiveness and reliability.

Conceptually, the quad-copter acts as a precision positioning tool, exploiting vertical access and low-altitude manoeuvrability to approach areas that are difficult or dangerous for ground forces to reach directly. Compared to direct-fire anti-armour weapons, which require line-of-sight and expose operators to immediate counter-fire, drones allow engagement from concealment and at stand-off distances. This has been particularly relevant in dense urban terrain, trench networks, and forested areas, where armoured vehicles and bunkers may be partially shielded by terrain or structures. However, the same low-altitude flight profile that enables this access also makes quad-copters vulnerable to weather, small-arms fire, electronic interference, and limited payload margins. Ref: Appendix (B) - Battalion based Drone Combat Unit.

When used against armoured personnel carriers, the tactical logic has focused on exploiting inherent design vulnerabilities rather than defeating armour through sheer force. Roof sections, engine compartments, and access hatches are generally less protected than frontal or side armour, and the aerial approach alters the geometry of the engagement. Even so, the effectiveness of such attacks is highly dependent on accuracy, timing, and the ability of the drone to remain stable while carrying a heavy and aerodynamically awkward payload. The addition of a rocket-propelled munition significantly reduces flight time, agility, and safety margins, increasing the likelihood of mission failure before reaching the target.

Against hardened bunkers and field fortifications, quad-copter-delivered munitions have been used to bypass the protective assumptions built into static defences. Bunkers are typically designed to resist horizontal fire and fragmentation, relying on earth cover, reinforced concrete, and narrow firing apertures. An aerially delivered munition can negate some of these advantages by entering from above or through openings never intended as points of attack. Yet even here, the physical limits of the drone impose constraints: payload weight reduces endurance, and the blast effects of small munitions may be insufficient to cause structural collapse, relying instead on shock, fragmentation, and psychological impact to neutralise occupants.

One of the most significant limitations of this tactic is power and range. Carrying heavy REG munitions often requires larger batteries or additional power reserves to maintain lift and control, which in turn adds weight and further reduces range. As a result, these drones typically operate over short distances and limited flight times, forcing operators to position themselves relatively close to the front line. This proximity increases exposure to detection, counter-battery fire, and electronic warfare. Battery constraints also limit loiter time, reducing opportunities for target selection and increasing pressure to act quickly once airborne.

Mobility and tempo remain essential to survival, but they also highlight another weakness. Drone launches generate electronic, acoustic, and visual signatures, and repeated use from the same area can quickly draw enemy attention. Units employing these systems must rely on rapid deployment, brief missions, and immediate displacement, which demands discipline, coordination, and reliable transport. Any delay—caused by battery changes, technical faults, or environmental conditions—can negate the advantages gained through surprise.

Despite these limitations, the psychological and doctrinal effects of quad-copter-delivered munitions are substantial. Armoured crews and infantry in fixed positions must contend with a persistent overhead threat that is difficult to predict and not easily countered by traditional means. This erodes confidence, restricts movement, and increases cognitive strain, even when actual physical damage is limited. In this sense, the drone functions as a tool of pressure and disruption as much as a weapon.

Overall, the use of quad-copters to carry rocket-propelled munitions illustrates how modern conflicts blend improvisation with established weapon concepts. While constrained by payload, battery life, range, and vulnerability to countermeasures, these systems demonstrate how adaptability, vertical access, and psychological dominance can allow small actors to challenge armoured vehicles and hardened positions, forcing a reassessment of protection, mobility, and survivability in contemporary warfare.

Scenario - 3 Offensive Tactics (Stealth against High Value Targets)

The use of low-cost UAV's to neutralise far more expensive armoured platforms illustrates a profound shift in the economics of modern warfare. A quad-copter costing in the order of USD 1,200 occupies the opposite end of the cost and complexity spectrum from a main battle tank valued at several million dollars, yet the exchange ratio strongly favours the cheaper system when the objective is functional denial rather than total destruction. Modern armoured vehicles depend on mobility, power generation, cooling, and electronics to remain combat-effective; once these subsystems are compromised, even an intact hull can become a stranded liability. This reality has forced armoured forces to reconsider how they manoeuvre and protect themselves in environments saturated with unmanned systems.

The deployment of small aerial drones to deliver limited but precisely placed thermite charge against vulnerable areas—such as engine compartments or exposed auxiliary systems—has repeatedly proven sufficient to render static vehicles unusable. These techniques began to emerge shortly after the full-scale invasion of Ukraine in 2022 and rapidly evolved into more sophisticated operations. Over time, missions became increasingly planned, often conducted at night, extended over greater distances, and supported by improvised staging or recharging points. What initially appeared as ad hoc innovation developed into repeatable, cost-effective methods that leveraged persistence and access rather than overwhelming force. In this context, a drone costing little more than consumer electronics could neutralise an armoured or artillery asset worth millions, reshaping the risk calculus on both sides.

From an analytical perspective, this trend highlights how unmanned systems challenge long-standing assumptions about depth and rear-area security. Darkness, terrain masking, and the absence of onboard human risk reduce many of the natural deterrents that historically protected logistics nodes, maintenance areas, and reserve formations far from the front line. Distributed launch points and forward staging further complicate defensive planning, enabling small teams to project effects deep into areas once considered relatively safe. Persistence and reach—rather than sheer firepower—have thus become defining characteristics of contemporary unmanned warfare, blurring the traditional distinction between "front" and "rear."

The broader significance lies less in any single drone platform than in the system-of-systems logic underpinning its use. Commercially derived drones benefit from rapid innovation cycles, global supply chains, and continuous improvements in batteries, sensors, navigation, and control software driven by civilian markets. Even incremental advances in energy density, propulsion efficiency, or signal resilience can yield disproportionately large operational effects when applied at scale. By contrast, armoured vehicles evolve slowly, are expensive to upgrade, and are constrained by weight, logistics, and entrenched doctrine. This asymmetry allows relatively unsophisticated tools to impose complex and costly defensive burdens on technologically advanced platforms.

Equally important is the counter-response. The growing reliance on electronic warfare, adapted air-defence measures, physical protection kits, and layered security reflects an acknowledgement that inexpensive drones are not temporary anomalies but enduring features of the battlespace. As countermeasures improve, unmanned platforms diversify in form and employment, reinforcing a cycle of adaptation rather than a return to pre-drone norms. Survivability increasingly depends on detection, integration, and resilience across the force, not solely on armour thickness or weapon calibre.

In strategic terms, the disparity between a low-cost drone and a multi-million-dollar tank symbolises a wider transformation in warfare. Power is no longer measured only by the possession of the most expensive platforms, but by how effectively forces combine affordability, adaptability, and innovation to impose costs on an adversary. This dynamic is likely to persist beyond the current conflict, shaping military planning, procurement, and doctrine worldwide.

Scenario - 4 Offensive Tactics (Quad-Copter-Tethered FPV) or (Fibre Linked)

The operational use of tethered quad-copters represents a notable evolution in unmanned aerial systems, shaped largely by the increasing density of electronic warfare and electromagnetic interference on modern battlefields. While early military drone concepts emphasised mobility, autonomy, and wireless control, the realities of contested electromagnetic environments have revived interest in a far older principle: the physical tether. By combining quad-copters with modern fibre-optic cabling, militaries and non-state actors alike have found ways to extend persistence, reliability, and control in situations where conventional radio-linked drones struggle to survive.

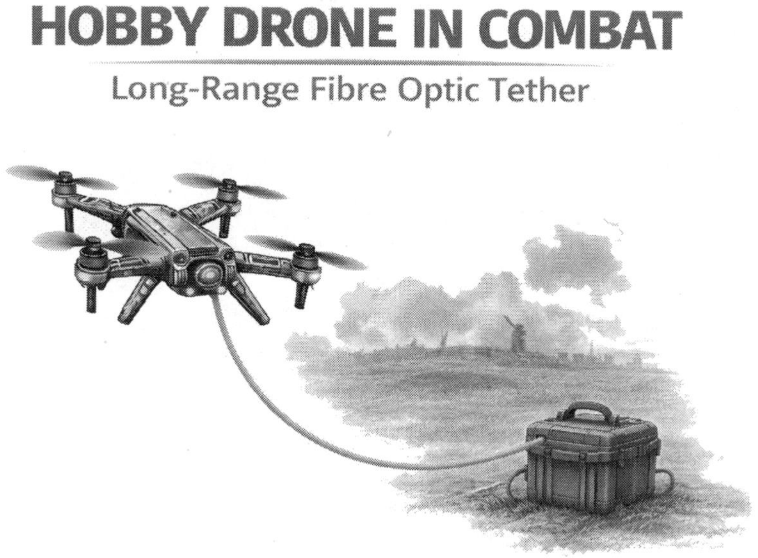

Historically, tethered aerial platforms are not new. Observation balloons and tethered aerostats were used extensively from the late nineteenth century through the First World War for reconnaissance and artillery spotting. What distinguishes contemporary tethered quad-copters is the convergence of lightweight electric propulsion, compact sensors, and ultra-thin fibre-optic cables capable of carrying high-bandwidth data over considerable distances. This technological convergence has allowed small drones to remain airborne for extended periods while maintaining secure, interference-resistant communication with ground operators.

One of the principal advantages of tethered quad-copters is their exceptional tolerance to electronic warfare. Because control signals and sensor data are transmitted through a physical fibre-optic link rather than radio frequencies, they are largely immune to jamming, spoofing, and interception. In operational environments saturated with electronic countermeasures, this resilience has proven decisive. Tethered drones can continue to provide stable video feeds, target observation, and situational awareness even when untethered drones are forced to land or are rendered ineffective. This capability has made them particularly valuable for static or semi-static roles such as perimeter surveillance, urban over-watch, hard kill missions and continuous monitoring of key terrain.

Beyond reconnaissance, tethered quad-copters have demonstrated adaptability across a wider range of missions. The reliable power and data link afforded by the tether enables longer on-station times and the use of heavier or more energy-intensive payloads than would otherwise be practical. As a result, tethered platforms have been employed not only for observation but also for communications relay, electronic sensing, and, in Ukraine, offensive roles. The physical link ensures precise operator control through FPV head-set and reduces the uncertainty associated with autonomous or intermittently connected systems, reinforcing their appeal in high-risk environments.

Recent advances in fibre-optic technology have further expanded the effective range of tethered drones. Modern cables are lighter, stronger, and capable of transmitting vast amounts of data over kilometres while imposing relatively modest weight penalties. This has allowed tethered quad-copters to operate at greater distances and altitudes than earlier generations, blurring the distinction between tethered and free-flying systems. In some operational contexts, these drones have achieved levels of persistence and reach that rival or exceed those of battery-limited untethered platforms.

However, the widespread operational use of fibre-optic-tethered drones has introduced a significant and often overlooked drawback: environmental and logistical pollution. As drones manoeuvre, reposition, or are forced to disengage under pressure, lengths of fibre-optic cable are left behind on the ground. Over time, this has begun to result in extensive accumulations of cable across fields, urban areas, and natural environments. Unlike traditional battlefield debris, fibre-optic strands are difficult to see, slow to degrade, and challenging to remove, posing hazards to civilians, vehicles, wildlife, and post-conflict recovery efforts. Ref: Appendix (B) example, 5 Fibre optic pollution in the Donbas Oblast of Ukraine. The whole town pictured, covered in a spiders web of fibre optics threads, from one side of the city to the other.

This pollution also creates secondary operational concerns. Discarded cables can reveal past drone positions, compromise concealment, and complicate movement for friendly forces. In prolonged conflicts, the accumulation of fibre-optic debris becomes a tangible reminder that even low-cost, high-tech solutions carry long-term consequences beyond their immediate tactical benefits. The environmental impact, while rarely decisive in the short term, raises important questions about sustainability and responsibility in modern warfare.

In sum, tethered quad-copters have proven highly successful across a spectrum of operational roles, largely due to their resilience against electronic interference and the expanding capabilities enabled by modern fibre-optic technology. Their effectiveness illustrates how adaptation to electronic warfare has driven innovation in unmanned systems. Yet their growing use also highlights an enduring tension in military history: solutions that offer decisive short-term advantages often introduce new and complex problems that persist long after the fighting has moved on.

After the war in the Donbas, the cleanup of fibre-optic cables used for drone control will pose a complex and largely unprecedented environmental and logistical challenge. These cables, often deployed hastily and in large quantities across fields, forests, and urban zones, are typically left behind—cut, damaged, or buried during combat operations. Unlike traditional wartime debris, fiber optic lines are thin, difficult to detect, and can be spread across thousands of kilo-meters, often without any records of where they were laid. Clearing them will likely require extensive manual labour, as current technologies offer limited ways to locate and extract such fine material. This task is further complicated by the risk of un-exploded ordnance in areas where cables were deployed, posing dangers to cleanup crews. In rural regions, leftover fiber could interfere with farming equipment, reduce soil productivity, or entangle with new infrastructure projects. Social and economic recovery may also be hindered as fibre remnants create confusion during telecom reconstruction and contribute to long-term visual and plastic pollution. Because most battlefield-grade fibre optic materials are not recyclable, large volumes may end up in landfills unless special disposal programs are created. Without coordinated cleanup efforts, this pollution may persist for years, subtly but significantly affecting the environment, agriculture, and civilian life in the post-conflict Donbas.

Scenario- 5 Sinking the Russian Moskva (Swarm Tactics)

14 April 2022 - Sinking of the Russian warship Moskva, the flagship of the Black Sea Fleet

The attack on the cruiser Moskva can be described as a two-phase, unmanned-enabled strike sequence designed to overwhelm a high-value naval platform through saturation and cognitive overload.

Phase One

Phase One consisted of the coordinated employment of multiple unmanned aerial systems of differing types and profiles, operating in a swarm-like manner to saturate the ship's defensive sensors, fire-control radars, and command-and-control processes. This unmanned swarm forced the vessel into a reactive defensive posture, degrading situational awareness, exhausting tracking capacity, and compressing decision-making timelines.

Phase Two

Phase Two followed once defensive coherence had been disrupted, with two land-based Neptune anti-ship missiles launched from divergent axes at low altitude. Unable to effectively discriminate, prioritise, and engage the combined volume of unmanned and missile threats, the ship's air-defence systems failed, allowing the missiles to strike the vessel. The resulting damage led to catastrophic system failures and the eventual sinking of the ship. Doctrinally, this sequence illustrates how low-cost, expendable unmanned systems can be employed as a decisive shaping tool, enabling a small number of precision weapons to defeat a technologically superior naval combatant by collapsing its defensive decision cycle rather than through brute-force firepower alone.

From a UAV warfare and military innovation perspective, the loss of the Russian cruiser Moskva can be analysed as an early illustration of swarm-enabled shaping operations preceding a kinetic missile strike. In this model, unmanned aerial systems were employed in coordinated numbers to saturate the ship's sensor and air-defence architecture, forcing the defender into a reactive posture and degrading its ability to maintain a coherent air picture. By overwhelming radar tracking capacity, distracting fire-control channels, and compressing decision timelines, the swarm functioned as a force multiplier rather than a direct strike asset. Once the ship's defensive systems were cognitively and technically saturated, Ukrainian forces launched two low-altitude anti-ship missiles from divergent axes, exploiting gaps created by the preceding unmanned assault. Doctrinally, this sequence demonstrates the growing effectiveness of distributed, low-cost, expendable drones used in swarming formations to exhaust and confuse high-value naval platforms, enabling relatively few precision weapons to achieve decisive effects. The case highlights a fundamental shift in maritime warfare: dominance is no longer determined solely by missile count or platform tonnage, but by the ability to integrate autonomous and semi-autonomous systems to collapse an adversary's defensive decision-making cycle prior to lethal engagement.

Scenario's- 6 Offensive Naval Tactics (Stealth High Value Targets)

September - 2023 The Boyko Towers

Since the expansion of the maritime dimension of the war, Ukrainian operations against Russian-occupied offshore energy platforms in the Black Sea have become a defining example of asymmetric warfare applied at sea, combining intelligence, special operations, and an increasingly sophisticated ecosystem of unmanned systems, with a clear sequence of missions unfolding over time. These installations—often referred to collectively as the "Boyko Towers"—were originally civilian gas-drilling platforms, but following their seizure they were converted into militarised outposts used for surveillance, electronic warfare, and force projection, making their isolation, fixed location, and symbolic value particularly attractive targets for Ukraine's intelligence and naval forces, notably the Main Directorate of Intelligence of Ukraine (GUR) and the Ukrainian Navy.

From the outset, drones played a central enabling role, beginning in early and mid-2023 with long-range reconnaissance UAVs tasked with mapping platform layouts, identifying radar, communications, and electronic warfare equipment, and monitoring Russian troop rotations and resupply activity. This preparatory phase set the conditions for the September 2023 mission, when Ukrainian military intelligence conducted what it described as a "unique operation" to temporarily recapture the Boyko Towers; during this raid, aerial reconnaissance drones and quad-copters provided continuous over-watch, relaying live video to commanders coordinating amphibious insertions, while FPV drones were held in reserve to suppress exposed positions. The operation resulted in the seizure of valuable equipment, including a Neva radar system, helicopter ammunition, and communications hardware, and reportedly coincided with an encounter involving a Russian Su-30 fighter jet, highlighting both the operational risk and the reliance on drones to extend situational awareness beyond the platforms themselves.

February 2024 "The Citadel" operation

As Russian forces further integrated the platforms into a wider Black Sea surveillance and targeting network through late 2023, Ukrainian drone use expanded in scale and complexity, leading into the February 2024 "Citadel" operation, during which Special Operations Forces seized key electronic systems—some used to enhance Russian drone control, signal relay, and maritime surveillance—before deliberately destroying the platform; throughout this mission, drones were used to coordinate timing, monitor enemy reactions, and conduct battle-damage assessment, ensuring both exploitation and denial objectives were achieved. The campaign continued with a August 2024 strike targeting a platform used for GPS spoofing, a form of electronic warfare that posed dangers not only to military systems but also to civilian maritime navigation; in this case, unmanned aerial systems were used to detect and confirm spoofing emissions, identify specific equipment, and support a precision strike that addressed both a tactical military threat and a broader maritime safety concern. The evolution of these tactics culminated in a mature, integrated approach to unmanned warfare, where drones functioned not merely as strike assets, but as persistent intelligence, surveillance, reconnaissance, and electronic warfare enablers—capable of shaping the maritime battle-space before, during, and after kinetic action.

November 2025 attack on the Syvash drilling platform,

November 2025 attack on the Syvash drilling platform, where drones were again central, detecting and tracking a Russian anti-tank missile crew and locating technical reconnaissance and observation systems prior to the strike, further degrading Russian situational awareness and force protection at sea. Across this timeline—from reconnaissance in early 2023, to raids and seizures in September 2023, platform destruction in February 2024, electronic warfare suppression in August 2024, and targeted strikes in November 2025—Ukrainian forces demonstrated an increasingly coherent operational system built around unmanned technologies: long-endurance UAVs for wide-area maritime surveillance, compact quad-copters and FPV drones for close-in reconnaissance and precision attack, and naval drones for stealthy approach and strike, all integrated with amphibious special forces when required. Strategically, this dated sequence of operations illustrates how Ukraine has challenged Russian naval dominance through asymmetry rather than parity, targeting high-value surveillance, electronic warfare, and command-and-control nodes far from the land front line, extending the battlefield into areas once considered secure and forcing Russia to divert resources into defence, dispersal, and counter-drone measures, while demonstrating that in modern maritime conflict, drones function not merely as weapons but as the connective tissue linking intelligence, precision attack, and special operations into a persistent and adaptive campaign. Reference, Appendix (B), example 5. Fibre Optic FPV with explosive ordinance.

Scenario's- 7 Fixed-Wing Reconnaissance (Development and Innovation)

2014 The Crimea Invasion

The development of long-range fixed-wing hobby-based drones in Ukraine can be traced back to the upheaval that followed the events of 2014, when the annexation of Crimea and the outbreak of fighting in the Donbas exposed severe gaps in Ukrainian reconnaissance and strike capabilities. At that time, Ukraine possessed limited indigenous unmanned aviation assets, and Western military assistance was cautious and incremental. In response, volunteers, engineers, and hobbyists began adapting commercially available fixed-wing model aircraft—originally intended for photography, mapping, or sport—into improvised reconnaissance platforms. These early systems were rudimentary, often assembled from foam air-frames with basic autopilots, but they provided vital aerial awareness over contested terrain.

2015 and 2016

Between 2015 and 2016, these grassroots efforts became more organised. Volunteer groups and informal workshops refined air-frame designs to improve endurance and range, recognising that fixed-wing aircraft offered significant advantages over multi-rotor platforms for long-distance flight. Drawing on civilian advances in GPS navigation, lightweight cameras, and lithium battery technology, Ukrainian operators began fielding drones capable of flying tens of kilometres, relaying imagery from beyond the immediate front-line. This period marked the transition from ad hoc experimentation to a recognisable ecosystem of semi-professional drone units supporting ground forces.

2017–2019

By 2017–2019, the use of fixed-wing drones had become institutionalised within parts of the Ukrainian military and affiliated volunteer networks. Hobbyist air-frames were increasingly replaced or supplemented by locally produced designs inspired by civilian unmanned aircraft. These drones were used primarily for reconnaissance and artillery spotting, reflecting a cautious doctrinal approach shaped by limited resources and the political sensitivities of the time. Nevertheless, the steady improvement in range, reliability, and autonomy during these years laid the technical foundations for later developments. Reference Appendix (B), example 4

February 2022 - full-scale Russian invasion

The full-scale Russian invasion in February 2022 marked a decisive turning point. Faced with an existential threat and an enemy possessing numerical superiority in conventional systems, Ukraine rapidly expanded and diversified its drone programmes. Fixed-wing platforms, many derived from or inspired by hobby aircraft, were adapted for longer-range missions that extended far beyond tactical reconnaissance. By mid-2022, reports emerged of Ukrainian drones reaching deep into occupied territory, signalling a shift toward strategic reach rather than purely local battlefield support. This evolution was driven by necessity: fixed-wing drones offered greater range, higher cruising efficiency, and the ability to bypass heavily contested front-line airspace.

2023 Countermeasures

Throughout 2023, the contest between Ukrainian drone developers and Russian countermeasures intensified markedly, with electronic warfare becoming increasingly pervasive across the battlefield. This escalation forced Ukrainian designers to rethink conventional approaches, driving experimentation in navigation resilience, autonomous flight profiles, and simplified control architectures capable of operating under severe signal degradation or complete communication loss. In this environment, fixed-wing drones proved particularly valuable, as their aerodynamic efficiency enabled extended-range missions even when control links were brief, intermittent, or deliberately denied. The influence of the civilian hobby sector remained clearly visible, with rapid prototyping, open-architecture systems, and continuous iteration allowing Ukrainian teams to adapt far faster than traditional military procurement and development cycles would normally permit.

During the same period, the use of hobby drone kits sourced from outside Ukraine increased significantly. These platforms offered an ideal foundation for further development, allowing designers to accelerate production, standardise components, and scale output quickly. Their widespread availability and cost-effective nature made them especially attractive in a high-attrition environment, enabling continuous refinement and deployment while reducing reliance on slow or constrained supply chains.

2024 and 2025 long-range drone capability

By 2024 and 2025, Ukraine's fixed-wing long-range drone capability had matured into a layered system encompassing reconnaissance, decoy, and strike-adjacent roles. While some platforms remained clearly derived from hobbyist origins, others represented fully indigenous designs informed by a decade of wartime experience. The line between "civilian" and "military" technology grew increasingly blurred, echoing earlier historical moments when industrial and improvised solutions converged under the pressure of prolonged conflict.

Looking toward 2026, historians are likely to view Ukraine's development of long-range fixed-wing drones as a case study in adaptive warfare. Beginning with improvised hobby aircraft in 2014, Ukraine transformed civilian technology into a strategic tool capable of operating across depth and distance. This progression illustrates how sustained conflict, constrained resources, and innovative communities can reshape the character of air power, reaffirming a recurring lesson of military history: that necessity, more than doctrine or budget, often drives the most consequential innovations.

CHAPTER 14

UAV Anti-Drone Shield Technologies

11.1.1 Anti-Drone Technologies Introduction

The rapid proliferation of unmanned aerial systems (UAS), commonly referred to as drones, has fundamentally reshaped the contemporary security environment, transitioning aerial capability from an exclusive state-controlled military asset to a widely accessible, commercially available, modular, and highly adaptable technology employed by non-state actors, criminal organisations, and hobbyists, thereby driving the emergence of Counter-Unmanned Aerial Systems (C-UAS) as a critical domain within defence, homeland security, maritime operations, and critical infrastructure protection.

Modern anti-drone systems are designed to detect, identify, track, and neutralise unauthorised or hostile drones while minimising collateral damage, legal exposure, and disruption to legitimate air traffic, a task complicated by the small size, low altitude, slow speed, and often non-metallic construction of contemporary UAVs that challenge traditional air-defence paradigms. Effective C-UAS architectures follow a strict kill-chain logic—detection, identification, tracking, decision, and neutralisation—where failure at any stage compromises system integrity, and must address a diverse threat spectrum ranging from micro and nano surveillance drones, through commercial quad-copters used for smuggling, ISR, or IED delivery, to high-threat FPV kamikaze drones, fixed-wing long-range strike UAVs, and extreme-risk swarming systems, as starkly demonstrated in modern conflicts such as the war in Ukraine.

Detection technologies form the foundation of C-UAS and include radar systems adapted with X-band and Ku-band frequencies, AESA architectures, and Doppler processing to detect low-RCS targets; passive RF sensors capable of identifying command links, telemetry, and video down-links; EO/IR systems providing positive visual and thermal confirmation for rules-of-engagement compliance; and acoustic arrays effective at short range in urban environments. Identification and classification increasingly rely on AI-driven sensor fusion to prevent fratricide, civil aviation interference, and legal violations. Neutralisation methods are broadly divided into soft-kill techniques—such as RF and GNSS jamming, spoofing, and cyber takeover—and hard-kill solutions including kinetic interceptors, net-based capture systems, and directed-energy weapons, each with distinct trade-offs in effectiveness, cost, and collateral risk. Modern defence doctrine therefore favours integrated, layered C-UAS systems combining long-range detection, mid-range electronic or directed-energy disruption, and close-in defeat options under unified command-and-control software. These systems are now standard across military formations, naval vessels, and forward operating bases, while civilian and critical-infrastructure deployments prioritise soft-kill approaches due to legal and ethical constraints. Maritime environments introduce additional complexity through sea clutter, horizon limitations, and the convergence of aerial and surface drones, necessitating integration with electronic warfare suites and close-in weapon systems. Legal and ethical considerations—encompassing airspace sovereignty, civil liberties, and electromagnetic spectrum regulation—continue to restrict active countermeasures primarily to state authorities, even as future developments point toward increased autonomy, AI-driven response loops, swarm-on-swarm defence, and low-cost attributable interceptors, underscoring that no single technology is sufficient and that only layered, integrated, and legally compliant C-UAS solutions can meet the accelerating pace of drone-enabled threats.

11.1.2 Classes of Threat Drones

Drones exhibit wide variation in physical scale, system complexity, and operational intent, resulting in markedly different threat profiles across classes. Micro and nano UAVs are typically employed for close-range espionage and surveillance tasks and generally present a low to medium threat due to limited payload capacity and range, whereas commercially available quad-copters represent a more serious medium-level threat given their ability to conduct smuggling operations, intelligence, surveillance, and reconnaissance (ISR), and, in some cases, deliver improvised explosive devices (IEDs). At the higher end of the threat spectrum, first-person-view (FPV) kamikaze drones are optimised for direct attack missions, combining low cost with high lethality and precision, while fixed-wing UAVs pose a similarly high threat through their capacity for long-range ISR and strike missions with extended endurance. The most severe category comprises swarming drones, which employ coordinated, networked behaviour to execute saturation attacks that can overwhelm defences, creating an extreme threat level due to their scalability, redundancy, and ability to defeat traditional single-target countermeasures.

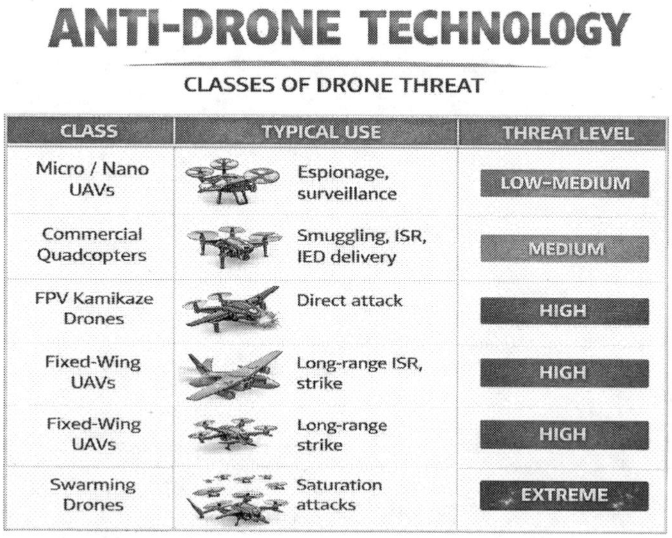

Modern conflicts—particularly the war in Ukraine—have demonstrated the lethality and cost-effectiveness of FPV drones and loitering munitions, forcing defenders to adopt low-cost, scalable countermeasures.

11.1.3 Core Principles of Anti-Drone Systems

The core principles of anti-drone systems, formally known as Counter-Unmanned Aerial Systems (C-UAS), are founded on a structured operational logic commonly referred to as the kill-chain model. This model reflects a sequential yet interdependent process designed to counter the full lifecycle of an unmanned aerial threat, from initial appearance in contested airspace through to its defeat or disruption. While individual technologies and doctrinal implementations vary widely between military, civilian, and maritime environments, all effective C-UAS architectures adhere to this fundamental framework. The kill chain consists of five critical stages—detection, identification, tracking, decision, and neutralisation—each of which must function reliably and in coordination with the others. Failure or degradation at any single stage undermines the effectiveness of the entire system, often rendering even the most advanced countermeasures ineffective. Consequently, modern anti-drone systems are designed not as isolated technologies but as integrated, layered ecosystems optimised to sustain performance across all phases of the kill chain under highly dynamic operational conditions.

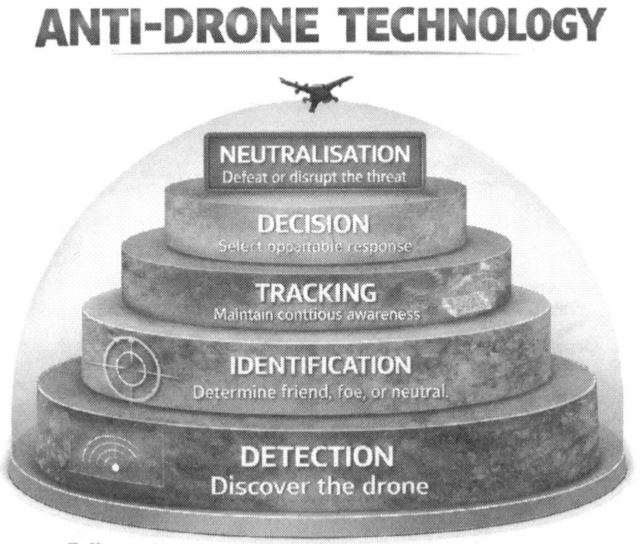

Failure at any stage compromises the entire system.

Detection represents the foundational stage of any anti-drone system and is universally recognised as the most critical and technically challenging element of the kill chain. Without timely and reliable detection, all subsequent stages become irrelevant. Detection involves the initial discovery of a drone within a defined volume of airspace, often before the operator or autonomous system has committed to its terminal mission. Unlike traditional aircraft, drones present a uniquely difficult detection problem due to their small physical size, low radar cross-section, slow speeds, and ability to fly at very low altitudes where ground clutter, terrain masking, and urban structures obscure sensor performance. Many small unmanned systems are constructed using plastics, composites, and lightweight alloys, further reducing their detectability by conventional air-defence radars designed to track fast-moving, metallic targets at medium to high altitude. As a result, modern C-UAS detection architectures rely on a combination of specialised sensors rather than a single modality, accepting that no individual sensor type can provide comprehensive coverage across all environments and threat classes.

Radar remains a cornerstone of drone detection, but its role has evolved significantly from traditional air-defence applications. Modern anti-drone radars operate in higher frequency bands, such as X-band and Ku-band, which provide improved resolution against small targets. Advanced signal processing techniques, including Doppler analysis and micro-Doppler signatures, are used to distinguish drone propeller movement from birds, debris, or environmental noise. These radars are often optimised for short-range, low-altitude coverage rather than long-range surveillance, reflecting the operational reality that most drone threats emerge suddenly and at close proximity to the defended asset. However, radar alone remains insufficient, particularly in complex terrain or dense urban environments where reflections and clutter degrade performance. This limitation has driven the widespread adoption of complementary detection technologies, most notably radio frequency sensing.

Radio frequency detection systems passively monitor the electromagnetic spectrum to identify emissions associated with drone operations, including command-and-control links, telemetry data, and video down-links. Because RF sensors do not emit energy themselves, they are inherently covert and immune to many forms of electronic counter-countermeasures. RF detection can often identify a drone before it is visible to radar, particularly when the operator initiates a control link at extended range. Additionally, RF systems can sometimes geolocate both the drone and its operator, providing valuable intelligence beyond immediate threat mitigation. Nevertheless, RF detection is inherently dependent on the drone emitting signals; fully autonomous drones, pre-programmed way-point systems, and fibre-optic-guided platforms may remain entirely silent in the RF spectrum, rendering this method ineffective against a growing subset of advanced threats. As adversaries increasingly adopt autonomous flight profiles to defeat jamming and detection, RF sensing must be integrated with other detection modalities rather than relied upon in isolation.

Electro-optical and infrared sensors provide another essential layer within the detection stage, offering visual and thermal confirmation of aerial objects. Electro-optical cameras enable high-resolution imaging during daylight conditions, allowing operators or automated systems to visually confirm the presence of a drone and assess its configuration, payload, and behaviour. Infrared sensors detect heat signatures generated by electric motors, batteries, and power electronics, making them particularly effective during night operations or in reduced-visibility conditions. EO/IR systems are indispensable for confirming detections generated by radar or RF sensors, reducing false alarms and ensuring compliance with rules of engagement in civilian and mixed-use airspace. However, EO/IR sensors are line-of-sight systems with limited range and are affected by weather, smoke, fog, and visual clutter. As such, they are best employed as confirmation tools rather than primary wide-area detection sensors.

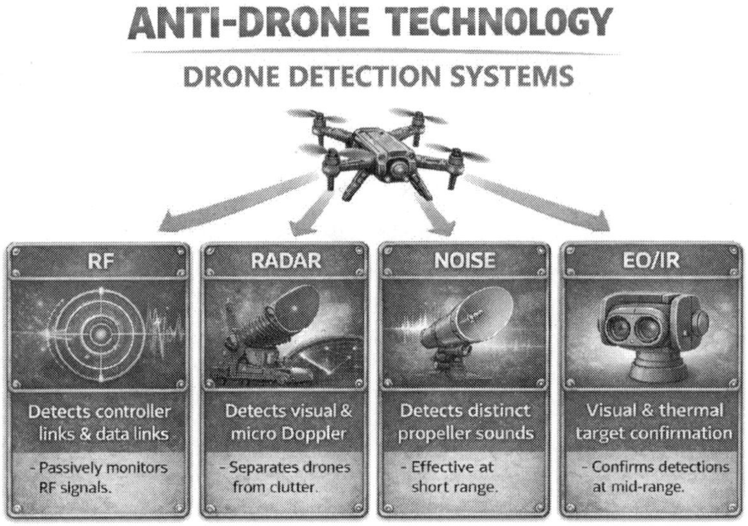

Acoustic detection systems, though more limited in range, play a valuable role in specific environments, particularly urban areas where buildings can shield radar coverage and RF signals may be congested. Acoustic arrays detect the distinctive sound signatures produced by drone propellers and motors, using pattern recognition algorithms to differentiate them from ambient noise. While acoustic sensors are highly sensitive at short range, their effectiveness degrades rapidly in high-wind conditions, noisy environments, or against fixed-wing drones operating at altitude. Despite these limitations, acoustic detection contributes to layered situational awareness and can provide early warning in scenarios where other sensors are obstructed.

Once a potential drone has been detected, the next critical stage in the kill chain is identification. Identification involves determining whether the detected object is a legitimate, authorised aerial platform or a hostile or unauthorised drone requiring intervention. This stage is essential for preventing fratricide, avoiding disruption to friendly operations, and ensuring legal and regulatory compliance, particularly in civilian airspace. The proliferation of legitimate drones used for commercial delivery, infrastructure inspection, agriculture, emergency services, and media has significantly increased the complexity of identification. A detection alone does not justify engagement; the system must reliably classify the object and assess intent before proceeding further in the kill chain.

Modern identification processes rely heavily on sensor fusion and data correlation. Radar signatures, RF emissions, visual characteristics, flight behaviour, and contextual information such as geofencing databases and airspace authorisations are combined to build a comprehensive threat profile. Artificial intelligence and machine learning algorithms are increasingly employed to classify drones based on patterns observed across multiple sensor inputs, enabling faster and more accurate identification than manual analysis alone. Identification systems may also integrate friendly drone registries and transponder data, allowing authorised platforms to be automatically recognised and excluded from further action. In military environments, identification may extend to assessing payload indicators, formation behaviour, or flight profiles consistent with reconnaissance, attack, or loitering missions.

Accurate identification is not merely a technical challenge but a strategic and legal necessity. Misidentification can lead to the neutralisation of friendly or civilian drones, causing operational disruption, economic loss, or escalation of conflict. Conversely, overly conservative identification thresholds may delay response until the threat is within lethal range. Effective C-UAS systems therefore balance speed and certainty, often incorporating graduated response options that allow for continued monitoring or warning measures prior to engagement. The reliability of identification directly influences the confidence and effectiveness of subsequent tracking and decision-making stages.

Tracking represents the continuous maintenance of situational awareness once a drone has been detected and identified. Unlike detection, which may occur intermittently, tracking requires persistent sensor coverage to monitor the drone's position, velocity, altitude, and trajectory over time. Effective tracking enables prediction of the drone's intent, identification of potential targets, and assessment of whether the threat is escalating or disengaging. Tracking is particularly challenging against agile drones capable of rapid manoeuvres, terrain masking, or sudden changes in speed and direction. Small UAVs can exploit clutter, fly close to structures, or descend to extremely low altitudes, causing intermittent sensor loss that degrades tracking continuity.

To address these challenges, modern C-UAS systems employ multi-sensor tracking architectures in which radar, EO/IR, and RF data are fused to maintain track continuity even when one sensor temporarily loses contact. Advanced tracking algorithms compensate for sensor dropout by predicting likely flight paths based on aerodynamic constraints and observed behaviour. In networked systems, tracking data may be shared across multiple sensors and platforms, extending coverage and reducing blind spots. Persistent tracking is particularly critical for distinguishing between transient overflight and deliberate attack, enabling defenders to allocate resources efficiently and avoid unnecessary engagement.

Tracking also plays a vital role in managing swarming threats, where multiple drones operate simultaneously, often with coordinated or semi-autonomous behaviour. In such scenarios, the ability to track individual drones within a swarm, as well as the collective behaviour of the group, becomes essential for prioritising targets and selecting appropriate countermeasures. Loss of tracking in swarm scenarios can rapidly overwhelm defensive systems, underscoring the importance of robust, scalable tracking architectures.

The decision stage represents the cognitive and command-and-control core of the anti-drone kill chain. At this stage, information gathered through detection, identification, and tracking is assessed to determine the most appropriate response. Decision-making may be performed by human operators, automated systems, or a hybrid combination depending on the operational context, and time constraints. The decision process must account for threat level, proximity to defended assets, rules of engagement, collateral risk, available countermeasures, and potential escalation consequences. In high-tempo military environments, decision timelines may be measured in seconds, whereas civilian infrastructure protection may permit longer deliberation and graduated responses.

Command-and-control systems play a central role in the decision stage, integrating sensor data into a coherent operational picture and presenting actionable options to decision-makers. Advanced C-UAS platforms increasingly incorporate decision-support algorithms that recommend responses based on predefined threat criteria and historical data. In some systems, particularly those designed to counter high-speed or swarming threats, automated engagement may be authorised within strict parameters to reduce reaction time. However, full autonomy in decision-making maybe constrained by legal, ethical, and political considerations.

The quality of decision-making is directly dependent on the integrity of earlier stages in the kill chain. Incomplete detection, ambiguous identification, or degraded tracking can lead to inappropriate responses, whether overly aggressive or insufficiently timely. As such, robust decision processes emphasise redundancy, validation, and adaptability, ensuring that decisions remain sound even under conditions of uncertainty or sensor degradation.

Neutralisation is the final stage of the anti-drone kill chain and involves the physical or functional defeat of the threat. Neutralisation methods are broadly categorised into soft-kill and hard-kill techniques, each with distinct advantages, limitations, and suitability depending on the operational context. Soft-kill approaches aim to disrupt the drone's ability to function without physically destroying it, typically through electronic or cyber means. Hard-kill methods involve the physical destruction or capture of the drone using kinetic or directed-energy weapons.

Soft-kill neutralisation commonly includes radio frequency jamming, which disrupts the command-and-control link between the drone and its operator, often causing the drone to hover, land, return to its launch point, or crash depending on its fail-safe programming. Global Navigation Satellite System jamming or spoofing interferes with the drone's navigation capability, causing loss of positional awareness or deliberate misguidance. Cyber takeover techniques exploit vulnerabilities in drone firmware or communication protocols to assume control of the platform. Soft-kill methods are particularly attractive in civilian environments due to their lower risk of physical collateral damage, but they are increasingly challenged by autonomous drones designed to operate without external signals.

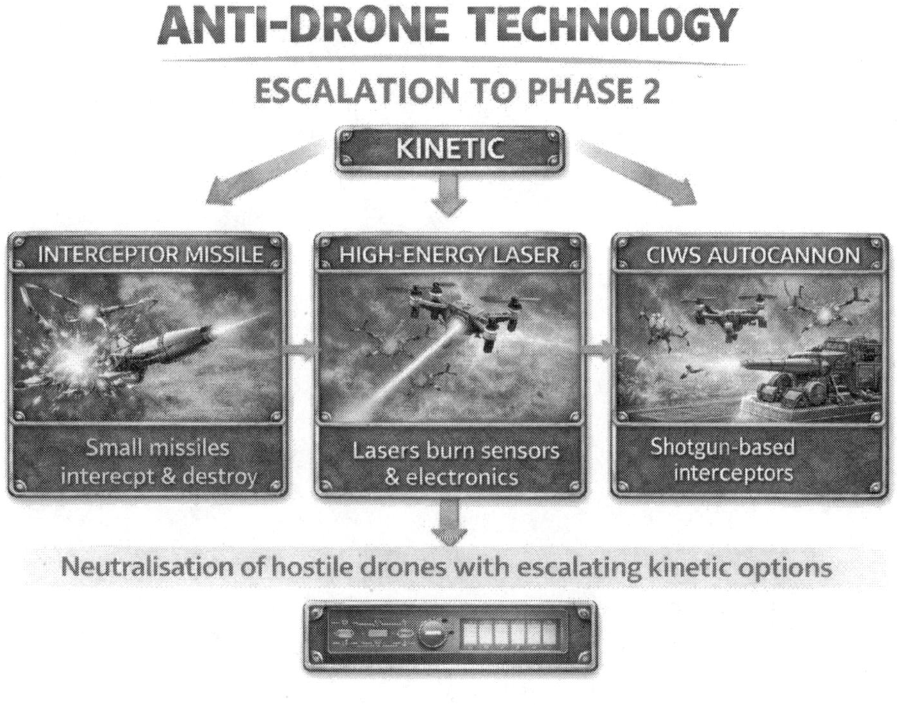

Hard-kill neutralisation methods include small missiles, auto-cannons, shotgun-based systems, interceptor drones, nets, lasers, and high-power microwave weapons. Kinetic solutions offer high certainty of defeat but carry significant risks in populated areas due to falling debris and missed shots. Directed-energy weapons provide rapid engagement, deep magazines, and low cost per shot, but require precise tracking and are affected by atmospheric conditions. Capture-based systems, such as nets deployed from interceptor drones, are well suited to urban and sensitive environments but are limited in range and scalability.

Effective neutralisation is not solely about defeating the drone but doing so in a manner consistent with mission objectives, and safety considerations. A successful engagement that causes unintended harm may be strategically counterproductive, particularly in civilian or humanitarian contexts. Therefore, modern C-UAS systems prioritise proportionality, selectivity, and integration, ensuring that neutralisation options are aligned with the threat profile and operational environment.

In summary, the core principles of anti-drone systems are inseparable from the kill-chain model that governs their design and operation. Detection establishes awareness, identification provides clarity, tracking maintains continuity, decision enables control, and neutralisation delivers effect. Each stage is dependent on the performance of the others, and weakness at any point compromises the entire system. As drone technology continues to evolve, particularly with the rise of autonomous, low-cost, and swarming platforms, the importance of integrated, layered, and resilient C-UAS architectures will only increase. The effectiveness of future anti-drone systems will be determined not by individual technologies, but by how well they sustain the integrity of the kill chain under conditions of complexity, speed, and uncertainty.

The six-layer, dome-style anti-drone architecture

The next diagram illustrates a six-layer, dome-style anti-drone and air-defence architecture, designed to defeat aerial threats through depth, redundancy, and escalating capability as targets approach protected assets.

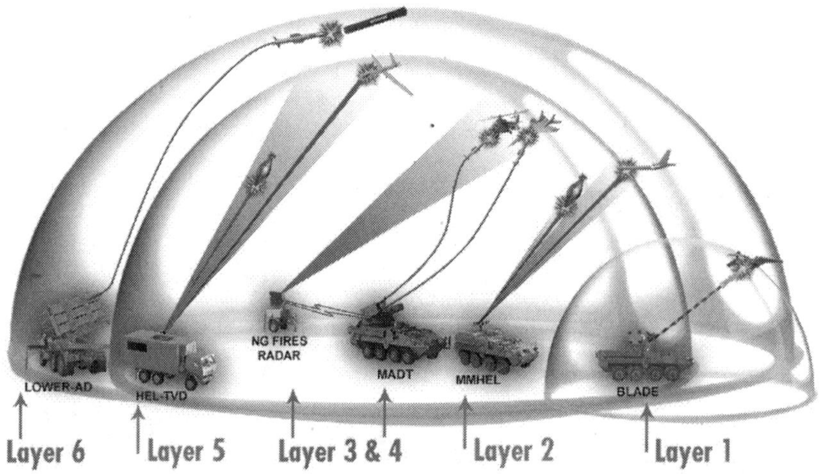

Layer 1 – Point Defence / Terminal Protection (BLADE):

Layer 1 is the innermost and final defensive barrier, responsible for protecting critical assets at extremely close range. Systems such as BLADE are typically highly mobile, vehicle-mounted point-defence platforms equipped with rapid-fire guns, short-range missiles, or directed-energy weapons. Their primary role is to intercept drones, loitering munitions, or small UAVs that have evaded all outer layers. Engagement ranges are short, reaction times are measured in seconds, and systems are optimised for high agility targets flying low, slow, or unpredictably. This layer accepts that some threats will leak through and focuses on last-second destruction to prevent direct impact on defended forces or infrastructure.

Layer 2 – Short-Range Air Defence (MMHEL):

Layer 2 expands the protective bubble outward using short-range air-defence (SHORAD) systems such as MMHEL. These platforms are designed to counter low-altitude drones, cruise-like UAVs, and loitering munitions before they reach the terminal zone. Equipment in this layer may combine missiles, cannons, electronic warfare payloads, or high-energy lasers. The emphasis is on rapid target acquisition, high engagement rates, and the ability to deal with swarms or multiple simultaneous threats. Layer 2 provides critical breathing space for inner defences by thinning incoming attacks and disrupting coordinated drone assaults.

Layers 3 & 4 – Mid-Range Integrated Air Defence (MADT + NG Fires Radar):

Layers 3 and 4 form the core of the defensive system, integrating medium-range interceptors such as MADT vehicles with Next-Generation Fires Radar and networked command-and-control. This layer is responsible for detecting, classifying, tracking, and engaging threats at tactically significant distances. Advanced radar systems provide persistent surveillance, cue shooters, and enable cooperative engagements across multiple platforms. Interceptors in this zone target a wide spectrum of threats, including fixed-wing UAVs, rotary drones, manned aircraft, and some missile classes. The strength of this layer lies in sensor fusion, data sharing, and coordinated fires, allowing multiple systems to act as a single defensive organism rather than isolated units.

Layer 5 – Long-Range Precision Intercept / High-Altitude (HEL-TVD):

Layer 5 introduces longer-range and higher-altitude engagement capabilities, such as HEL-TVD or similar advanced interceptor and directed-energy systems. This layer focuses on destroying or disabling threats before they can descend into lower airspace, where drone density and clutter increase. Equipment here may include high-energy lasers, long-range surface-to-air missiles, or specialised counter-UAS systems capable of engaging fast or high-flying targets. By neutralising threats early, Layer 5 significantly reduces the load on inner layers and limits the effectiveness of massed or coordinated drone attacks.

Layer 6 – Outer Air Defence / Early Warning (LOWER-AD):

Layer 6 represents the outermost defensive envelope, providing early warning, long-range detection, and initial engagement of hostile aircraft, missiles, and large UAVs. Systems such as LOWER-AD are designed to operate at extended ranges, often integrating with national or theatre-level air-defence networks. This layer is critical for shaping the battle-space: detecting threats shortly after launch, cueing lower layers, and destroying high-value or high-speed targets before they approach defended areas. Even when threats are not destroyed outright, Layer 6 degrades formations, forces evasive manoeuvres, and disrupts timing, making subsequent layers more effective.

Overall Concept:

Together, these six layers form a resilient, overlapping defensive dome that accepts no single system is sufficient on its own. Instead, survivability is achieved through layered coverage, mixed sensor types, diverse intercept methods, and escalating engagement ranges. The architecture is specifically suited to modern battlefields dominated by drones, swarms, electronic warfare, and asymmetric aerial threats, ensuring that even low-cost UAVs face a high probability of interception at multiple points along their attack path.

11.1.4 Drone Hard-Kill Options

Hard-kill neutralisation methods represent the "physics-based" end of the counter-UAS toolkit. Rather than denying a drone its communications or navigation links, as in soft-kill approaches, hard-kill systems aim to destroy the platform outright, disable it through physical damage, or forcibly remove it from the airspace. Their principal advantage lies in their effectiveness against autonomous drones, pre-programmed flight profiles, and fibre-optic-controlled systems that are increasingly resistant to RF and GNSS disruption. At the same time, hard-kill solutions introduce their own constraints, most notably the risks associated with collateral damage, falling debris, missed shots, and the requirement for reliable detection, identification, and tracking to avoid engaging friendly or non-hostile platforms. As a result, contemporary counter-drone doctrine rarely relies on a single hard-kill mechanism. Instead, it emphasises layered engagement zones in which guns, interceptors, and, where available, directed-energy weapons are combined to provide overlapping coverage, with defenders selecting the least hazardous and most certain option that still meets the required time-to-intercept.

Small missiles remain the archetypal hard-kill solution and continue to play a critical role against higher-value or more demanding aerial threats such as loitering munitions, larger fixed-wing UAVs, and high-speed one-way attack drones. Their defining strengths are long engagement range and a high probability of kill when integrated with capable sensors, guidance systems, and fire-control solutions. These attributes make missiles particularly valuable in the outer layers of a defensive system, where threats must be neutralised at distance before they approach protected assets. However, missiles are also characterised by high cost per engagement and limited magazine depth. In operational environments dominated by large numbers of inexpensive drones, this cost imbalance can rapidly become unsustainable, driving a shift toward cheaper defeat mechanisms for small, slow, or low-value targets. The resulting trend is not the obsolescence of missile-based defence, but its increasingly selective application, reserved for targets that exceed the practical reach or effectiveness of guns, interceptors, and point-defence systems, or where rules of engagement demand a high-confidence intercept with minimal dwell time over sensitive areas.

Auto-cannons and air-burst munitions have experienced a marked resurgence as counter-drone weapons, largely because they offer a favourable balance between effectiveness and cost when employed at scale. Medium-calibre guns, typically in the 20–35 mm range, are well suited to engaging small UAVs at short to medium ranges, particularly when paired with programmable ammunition. Airburst rounds detonate in proximity to the target, generating a cloud of fragments that compensates for the small size, erratic flight paths, and limited radar cross-section of drones. These systems are commonly mounted on vehicles or fixed platforms and integrated with radar, electro-optical, infrared, or fused sensor inputs that cue the weapon onto the target. Their flexibility allows deployment close to manoeuvre units, critical infrastructure, or high-value sites, where rapid reaction and sustained fire are essential.

Despite their effectiveness, auto-cannons are constrained by line-of-sight requirements, ammunition logistics, and the need for accurate target discrimination. Mis-identification risks fratricide or unintended harm, particularly in congested airspace or urban environments. Unlike missile systems, guns cannot independently search large volumes of airspace and are therefore heavily dependent on upstream sensors for timely and accurate cueing. If detection or tracking is degraded, the effectiveness of the gun system deteriorates rapidly. Nonetheless, when integrated into a broader sensor network, auto-cannons provide a robust and economical hard-kill layer capable of defeating a wide range of small to medium drone threats.

Shotgun-based systems appear, at first glance, incongruous in an era defined by networked sensors, autonomy, and artificial intelligence. Yet they have emerged as a practical point-defence solution in environments characterised by intense drone activity and short engagement ranges. Shotguns are portable, inexpensive, and can be fielded rapidly at the lowest tactical levels, providing a last-ditch defensive option when other measures have failed or are unavailable. Their limitations are clear: extremely short effective range, dependence on operator skill, and inconsistent results against fast or manoeuvring targets. Successful employment requires training, situational awareness, and favourable engagement geometry.

The operational logic behind shotgun use is straightforward. As drones proliferate and electronic countermeasures lose effectiveness against certain guidance modes, defenders require a method that remains viable even in the absence of sensors, networks, or electronic effects. Shotgun-based defence fulfils this role as a final layer of protection. Importantly, this approach is not limited to shoulder-fired weapons; experimentation has extended to mountable and drone-borne shotgun systems intended to engage hostile UAVs at ultra-close range. These adaptations reflect a broader pattern in counter-drone warfare: rapid, pragmatic innovation driven by immediate operational necessity rather than long development cycles.

Interceptor drones—unmanned systems designed specifically to destroy other drones—represent one of the most significant recent developments in hard-kill counter-UAS. Their primary impact lies in altering the cost dynamics of air defence. Instead of expending an expensive interceptor missile to defeat a low-cost UAV, defenders can deploy a comparatively inexpensive interceptor drone, often built from commercially derived components and produced at scale. This approach allows rapid iteration and adaptation as adversary tactics evolve, mirroring the flexibility of the drone threat itself.

Interceptor drones offer scalability advantages that traditional point-defence systems struggle to match. They can be launched from dispersed locations, repositioned dynamically, and integrated into distributed sensor networks that already support surveillance and FPV operations. Their limitations include vulnerability to electronic attack, dependence on operator skill where autonomy is limited, and the risk that engagements devolve into contests of numbers and training rather than technological superiority. Even so, the broader trajectory is clear: hard-kill counter-drone systems are becoming increasingly attributable and software-defined, borrowing production methods, command concepts, and operational tempo from the drone ecosystem they are designed to counter.

Nets and capture systems occupy a distinct niche within the hard-kill spectrum. Rather than destroying the drone, these systems physically entangle it or force a controlled capture. This approach is particularly valuable where explosive effects, fragmentation, or falling debris are unacceptable, such as in dense urban environments, around critical infrastructure, or in certain maritime contexts. Net-based solutions can be ground-launched, deployed from interceptor drones, or integrated into dedicated capture platforms. Their principal strengths are safety and intelligence value; captured drones can be examined to extract technical, tactical, and forensic information.

The limitations of net systems are primarily geometric and tactical. Effective engagement typically requires close range, favourable approach angles, and manageable target speeds. Nets are poorly suited to high-speed or highly manoeuvrable drones and can struggle to scale against multiple simultaneous threats. Nevertheless, within the inner defensive ring, net-based capture remains an important option when authorities require an outcome that is both tactically effective and legally defensible, prioritising control and accountability over outright destruction.

High-energy laser weapons are the most prominent directed-energy hard-kill option and have attracted sustained interest because of their potential advantages in cost per engagement and magazine depth. Once deployed, a laser system can theoretically engage large numbers of targets with minimal marginal cost, limited primarily by power generation and thermal management. Lasers also offer extremely rapid time-to-target, making them well suited to defeating small, fast-approaching drones within short engagement windows.

Latest interceptor drone styles for hard kill options

The constraints on laser systems are rooted in physics and engineering rather than concept. Atmospheric conditions such as fog, rain, dust, smoke, and turbulence can attenuate or scatter the beam, reducing effectiveness. Precision beam control and stabilisation are required to maintain focus on a small, moving target long enough to achieve a functional kill, whether through burn-through of structural components, sensor damage, or thermal failure of electronics. Accurate tracking and deconfliction are essential, as laser engagements are line-of-sight and intolerant of targeting error. Where environmental conditions and sensor integration are favourable, lasers are particularly well suited to persistent defence against repeated small attacks and may offer significant advantages against swarming threats if the system can transition rapidly between targets.

High-power microwave (HPM) weapons address a different set of challenges. Instead of concentrating energy on a single point, HPM systems emit electromagnetic energy across a broader area or cone, disrupting or damaging electronic components within multiple drones simultaneously. This makes them especially attractive for counter-swarm scenarios, where sequential hard-kill engagements risk being overwhelmed by numbers. The primary mechanism of effect involves upsetting, overloading, or permanently damaging onboard electronics, exploiting the inherent fragility of small UAV control, navigation, and power systems.

HPM systems introduce their own operational complexities. Effects can be difficult to predict precisely in dense electromagnetic environments, and there is potential risk to friendly electronics if emissions are not carefully managed. Like other directed-energy weapons, HPM systems demand substantial power, cooling, and integration with sensors and command-and-control architectures. Despite these challenges, their ability to achieve one-to-many effects without expending physical ammunition makes them increasingly attractive as drone swarms become more plausible and individual drones more disposable.

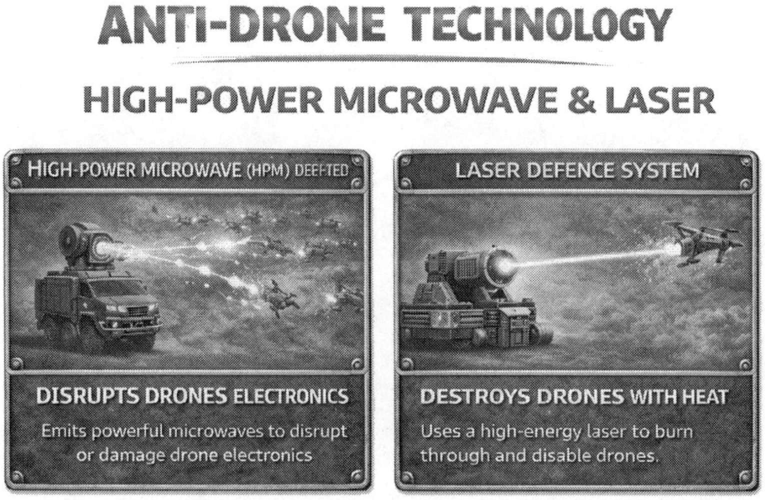

Latest interceptor drone styles for hard Kill options

The overarching lesson from the evolution of hard-kill counter-UAS methods is that no single weapon provides a universal solution. Effective defence emerges from the integration of complementary defeat mechanisms within a layered architecture. Missiles offer reach and high confidence against demanding targets; auto-cannons provide economical volume fire when properly cued; shotguns and nets serve as last-ditch or low-collateral options in the inner defensive ring; interceptor drones reshape the cost curve and enable rapid adaptation; and directed-energy systems promise deep magazines and, in the case of microwaves, area effects against multiple targets. The pace of drone innovation continues to compress the timeline between concept, fielding, and obsolescence, reinforcing a central truth of modern counter-drone warfare: the most effective hard-kill solutions are those that can be produced, trained, and sustained at the same speed as the threat they are designed to defeat.

Example, Directed Energy Weapon for Air Defence
Ukraine's "Sunray" Laser System February 2026

Within Phase 2 of the kinetic air defence kill chain—engagement and neutralisation—Ukraine has accelerated the development of improvised yet technologically sophisticated counter-drone systems. Among the most significant innovations is Sunray, a compact, vehicle-mounted directed-energy weapon engineered to defeat small unmanned aerial systems (sUAS) at dramatically reduced cost compared to traditional interceptor missiles.

Developed in roughly two years at a reported cost of only a few million dollars, and projected to retail for several hundred thousand per unit, Sunray reflects a wartime engineering model driven by urgency rather than protracted procurement cycles. The system is compact enough to be mounted on a civilian pickup platform, enhancing mobility and concealment in distributed battlefield environments.

11.1.5 Swarm, Interceptors

The rapid evolution of unmanned aerial systems has ensured that counter-UAS (C-UAS) is no longer a static defensive discipline but a continuously adapting technological contest, in which each advance in drone capability provokes an equally rapid countermeasure, driving an accelerating cycle of innovation. Emerging trends indicate that future C-UAS systems will increasingly depend on artificial intelligence, autonomous response loops, swarm-on-swarm defensive concepts, and low-cost attributable interceptors, all of which reflect a fundamental shift away from manpower-intensive, platform-centric air defence toward software-driven, distributed, and scalable architectures. Artificial intelligence is becoming central because the sheer volume, speed, and complexity of modern drone threats exceed the cognitive and reaction limits of human operators. Small drones are difficult to detect, can appear with minimal warning, and often operate in cluttered environments alongside civilian air traffic, birds, and ground interference.

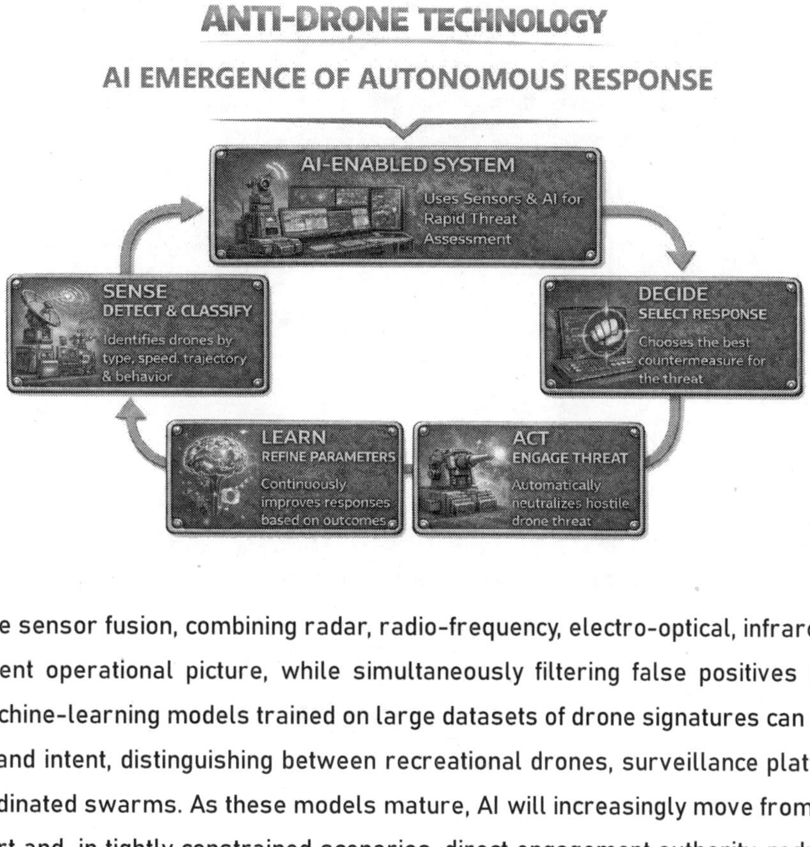

AI enables real-time sensor fusion, combining radar, radio-frequency, electro-optical, infrared, and acoustic inputs into a coherent operational picture, while simultaneously filtering false positives and prioritising genuine threats. Machine-learning models trained on large datasets of drone signatures can classify targets by type, behaviour, and intent, distinguishing between recreational drones, surveillance platforms, loitering munitions, and coordinated swarms. As these models mature, AI will increasingly move from advisory roles into decision-support and, in tightly constrained scenarios, direct engagement authority, reducing detection-to-response timelines from minutes to seconds.

Closely linked to AI is the emergence of autonomous response loops, which represent a decisive break from traditional command-and-control paradigms. In conventional air defence, human operators detect, identify, decide, and engage in a largely linear sequence, with each step introducing delay. Autonomous response loops compress or eliminate this latency by allowing the system to sense, decide, and act within predefined rules of engagement and safety constraints. Once a threat meets certain criteria—such as speed, trajectory, proximity to protected assets, or confirmed hostile behaviour—the system can automatically select the most appropriate countermeasure and execute an engagement without awaiting manual authorisation. This does not imply the removal of human oversight; rather, humans increasingly supervise the system at a higher level, setting engagement policies, ego-, and escalation thresholds, while the machine handles execution at machine speed. This approach is particularly critical against fast, low-altitude drones and saturation attacks, where human reaction times are insufficient. Autonomous loops also enable continuous adaptation, as AI models refine engagement parameters based on observed outcomes, improving effectiveness over time.

One of the most significant emerging concepts is swarm-on-swarm defence, which directly mirrors the offensive use of coordinated drone swarms. As attackers exploit numbers, redundancy, and decentralised control to overwhelm traditional defences, defenders are responding with their own cooperative unmanned systems. Defensive swarms can patrol airspace, share sensor data, distribute targets among themselves, and coordinate intercepts in a manner that is inherently scalable. Rather than relying on a single high-value sensor or weapon, swarm-based defence disperses capability across many inexpensive nodes, increasing resilience against jamming, kinetic attack, or system failure. In practice, swarm-on-swarm defence may involve interceptor drones that autonomously allocate targets, form barriers, herd hostile drones away from sensitive areas, or physically destroy them through ramming, entanglement, or onboard reflectors. The defensive swarm's strength lies not in the performance of any single drone, but in collective behaviour enabled by distributed AI and robust communications, allowing the system to continue functioning even when individual units are lost.

Low-cost attributable interceptors are a direct response to the economic realities of modern drone warfare and security operations. Traditional air defence systems were designed to counter high-value aircraft and missiles, making them ill-suited to defeating large numbers of inexpensive drones. Attributable interceptors invert this cost equation by accepting that defensive drones may be lost in the process of interception and designing them accordingly. These systems prioritise affordability, ease of manufacture, and rapid deployment over longevity and survivability. They are often built using commercial components, additive manufacturing techniques, and simplified air-frames, allowing them to be produced in large quantities and replaced quickly. Attributable interceptors may be single-use ramming drones, reusable net-capture platforms, or small kinetic interceptors designed for limited engagements. Their low cost enables defenders to match or exceed the attacker's numbers, restoring balance in scenarios where saturation would otherwise overwhelm fixed defences.

The convergence of AI, autonomy, swarm logic, and attributable platforms is transforming C-UAS from a static defensive layer into an adaptive, learning ecosystem. Future systems will not rely on a single sensor or weapon but will function as networks of sensors, reflectors, and decision nodes, dynamically re-configuring themselves in response to changing threats. For example, an AI-enabled C-UAS network may detect unusual RF activity, cue electro-optical sensors to confirm a drone's presence, predict its likely flight path using behavioural modelling, and dispatch a mix of interceptor drones and directed-energy systems based on range, weather conditions, and collateral-risk considerations. As the engagement unfolds, the system can update its predictions and reassign resources in real time, effectively "fighting the fight" faster than a human-managed system ever could.

Another important trend is the increasing emphasis on adaptability and software-defined capability. Rather than deploying fixed hardware solutions that require lengthy upgrade cycles, future C-UAS systems will rely heavily on software updates to introduce new detection algorithms, threat libraries, and engagement tactics. This allows defenders to respond quickly to adversary innovations such as new flight profiles, materials designed to reduce detectability, or novel guidance methods. Software-defined architectures also enable interoperability between different sensors and effectors, allowing systems from multiple vendors to operate within a single defensive framework. This flexibility is critical as no single technology can address all drone threats across all environments, and layered, integrated solutions are increasingly the norm.

The human role in future C-UAS operations will evolve rather than disappear. Operators will shift from direct control toward supervision, mission planning, and ethical oversight. As systems become more autonomous, ensuring compliance with legal and ethical constraints becomes a central design requirement. AI-driven decision-making should be transparent, audible, and constrained by clear rules of engagement to prevent unintended escalation or harm to non-combatants. Human-machine teaming concepts should therefore be integral, with humans retaining authority over escalation thresholds and system configuration while relying on automation for execution speed and precision.

Environmental and contextual awareness will also play a larger role in future developments. AI systems will increasingly incorporate terrain data, weather conditions, electromagnetic environment assessments, and civilian activity patterns into their threat evaluations. This enables more nuanced responses, such as selecting non-destructive interception methods in urban areas or prioritising stealthy tracking over immediate engagement when attribution and intelligence collection are more valuable than destruction. Such contextual intelligence reinforces the shift toward C-UAS as a holistic security function rather than a purely military capability.

The accelerating pace of innovation on both sides of the drone-counter-drone equation suggests that adaptability itself is becoming the most valuable capability. Offensive drone developers rapidly exploit commercial technology cycles, software updates, and modular payloads, while defenders must counter these advances without the luxury of long development time-lines. AI-driven learning systems, autonomous response loops, swarm-based defence, and attributable interceptors collectively address this challenge by enabling rapid iteration, scalability, and resilience. Rather than attempting to achieve permanent technological dominance, future C-UAS strategies will focus on maintaining a favourable balance through continuous adaptation.

Ultimately, the race between drone innovation and counter-drone adaptation is accelerating because the barriers to entry for drone development are low, while the potential impact of drone misuse is high. This dynamic ensures that C-UAS will remain an active and evolving field, shaped by advances in artificial intelligence, autonomy, and distributed systems. Success will depend less on any single breakthrough technology and more on the ability to integrate emerging tools into coherent, flexible, and ethically governed defensive ecosystems capable of evolving as fast as the threats they are designed to counter.

11.1.6 Summary, Anti Drone

Anti-drone technology has rapidly evolved from a niche military concern into a central pillar of modern security architecture, reflecting the profound transformation brought about by the widespread availability of unmanned aerial systems. Drones are no longer rare, state-controlled platforms; they are inexpensive, modular, and adaptable tools employed across military, civilian, commercial, and humanitarian environments. This ubiquity has fundamentally altered the threat landscape. Small, low-flying, and often autonomous drones can be used for reconnaissance, smuggling, disruption, coercion, or direct attack, frequently at a fraction of the cost of traditional aerospace systems. As a result, the ability to detect, identify, track, and defeat hostile or unauthorised drones has become an essential requirement for protecting people, infrastructure, and operations across an exceptionally broad range of contexts.

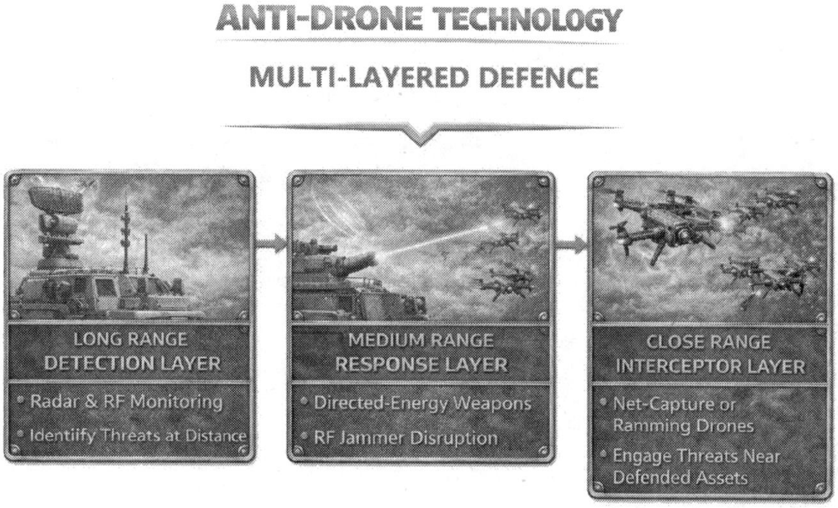

As outlined in section 11.2.3, at the core of effective counter-unmanned aerial systems (C-UAS) is the recognition that no single technology can address the drone threat in isolation. Drones vary enormously in size, speed, altitude, control method, and mission profile, from micro-UAVs conducting close-range surveillance to long-range fixed-wing systems and coordinated swarms. This diversity means that any one sensor or reflector will inevitably have blind spots or vulnerabilities. Consequently, modern anti-drone solutions are increasingly built around layered and integrated architectures that combine multiple detection methods, decision-support systems, and response options into a coherent whole. The emphasis has shifted from individual weapons or sensors to systems-of-systems designed to maintain resilience even when individual components are degraded or bypassed.

Detection and identification form the foundation of this layered approach. Traditional air-defence radars were not designed to detect small, slow, and low-altitude targets, prompting the development of specialised radar modes, radio-frequency monitoring, electro-optical and infrared sensors, and acoustic detection systems. Each has strengths and limitations: radar provides range and coverage but can struggle with clutter; RF detection is passive and effective against remotely piloted drones but less so against autonomous systems; EO/IR sensors offer positive visual identification but are constrained by weather and line-of-sight; acoustic systems can be valuable in urban settings but are vulnerable to environmental noise. Sensor fusion, increasingly supported by artificial intelligence, is therefore critical, allowing data from disparate sources to be combined into a single, more reliable operational picture.

Once a drone is detected and identified, decision-making becomes the decisive element. The challenge is not merely technical but operational, legal, and ethical. Operators must determine whether a drone is hostile, negligent, or benign; assess the potential consequences of action or inaction; and select an appropriate response within often stringent rules of engagement. This is particularly important in civilian and humanitarian contexts, such as airports, power infrastructure, refugee camps, or disaster-response zones, where the risk of collateral damage or interference with legitimate activity is high. Increasingly, decision support is augmented by automation and AI-assisted tools that can prioritise threats, recommend responses, and reduce reaction times, while still preserving human oversight where required.

Neutralisation options illustrate most clearly why layered solutions are essential. "Soft-kill" measures, such as RF jamming, GNSS denial, or cyber techniques, can be highly effective against many commercial and remotely piloted drones, often with minimal physical risk. However, they are ineffective against autonomous systems, hardened platforms, or drones using non-RF control methods. "Hard-kill" measures, including guns, missiles, interceptor drones, nets, lasers, and high-power microwave systems, provide physical defeat but introduce their own constraints related to cost, safety, scalability, and environmental conditions. The practical response has been to integrate multiple effectors into concentric engagement zones, selecting the least hazardous and most cost-effective option that can reliably defeat the threat at a given stage.

This layered philosophy is evident across military, civilian, and humanitarian domains, albeit with different emphases. In military environments, anti-drone systems are now integral to force protection, manoeuvre warfare, and air defence, countering not only reconnaissance drones but also loitering munitions and massed attacks. Mobility, rapid deployment, and survivability are critical, and systems must function under conditions of electronic warfare and physical attack. In civilian settings, the priority shifts toward safety, legality, and continuity of operations. Airports, prisons, ports, and critical infrastructure sites require systems that favour detection, identification, and non-destructive mitigation wherever possible, with kinetic options reserved for extreme circumstances. Humanitarian operations face a distinct challenge: drones can be both a threat and a valuable tool for aid delivery, mapping, and situational awareness, requiring nuanced policies and technologies that protect vulnerable populations without undermining legitimate use.

Legal and regulatory considerations are therefore inseparable from technical design. Airspace sovereignty, privacy rights, electromagnetic spectrum regulation, and the lawful use of force all constrain how anti-drone technologies can be deployed. In many jurisdictions, only state authorities are permitted to employ active countermeasures, and even then under tightly defined conditions. The most effective C-UAS systems are those designed with these constraints in mind, embedding compliance, audibility, and clear command authority into their architecture rather than treating legality as an afterthought.

Looking ahead, the trajectory of drone and counter-drone development suggests an accelerating cycle of adaptation. Advances in artificial intelligence, autonomy, and low-cost manufacturing are enabling increasingly capable drones, including coordinated swarms and systems designed to evade detection or resist electronic attack. In response, counter-drone systems are becoming more autonomous, more software-defined, and more reliant on scalable, attributable solutions such as interceptor drones and directed-energy weapons. The emphasis is shifting from exquisite, high-cost interceptors to sustainable defences that can match the pace, volume, and economics of the threat.

The central lesson is clear: anti-drone technology is no longer optional, nor can it be reduced to a single sensor or weapon. Effective defence against drones requires layered, integrated systems that combine diverse detection methods, intelligent decision-making, and a spectrum of response options, all operating within a robust legal and ethical framework. As drones continue to proliferate and evolve, the success of counter-drone efforts will depend less on any individual breakthrough and more on the ability to integrate technology, policy, and operations into coherent, adaptable solutions capable of protecting lives, infrastructure, and humanitarian space in an increasingly contested air domain.

CHAPTER 15

DIY Build, Project Hardware

12.1.1 Introduction to Project Build

This example project details the hardware required to build a standard, (typical) four-motor quad-copter drone as March 2026, along with the open-source software used for its configuration. It serves as a representative model of how a hobby drone could be built or modified for potential use in warfare. However, it is important to note that drones constructed to these specifications would not be legally operable in most countries across Europe, North America, and Asia. Current regulations in these regions typically mandate that drones operate within strict parameters, including altitude and range limits enforced through manufacturer-configured Geo-fencing. Additionally, operators are generally required to maintain visual line-of-sight with the drone at all times to ensure safe operation.

[IMPORTANT]

The following project example is provided to illustrate what can be constructed within the scope of a hobby project using readily available, off-the-shelf components sourced from local hobby shops or on-line retailers. This example is for illustrative purposes only. All users are responsible for ensuring that they operate drones in compliance with the laws and regulations applicable to their country or location, and for consulting the most up-to-date information for all components used before attempting to construct or operate a functional drone.

Please note that this project requires intermediate to advanced proficiency in electronics, software, and firmware development. It is not suitable for beginners, as it assumes prior experience with embedded systems, system integration, and troubleshooting at both the hardware and software levels. However, for individuals with an appropriate foundational background, the project provides a valuable practical introduction to the depth and breadth of knowledge required to undertake comparable advanced systems-development efforts in this field.

Hardware - List of Components

1. 1 x Frame, GEP-Mark4-7 Frame. Seven Inch Frame type H, carbon fibre light weight.
2. 1 x Flight & ESC Controller. SpeedyBee F405 V3 Stack BLS 50A. F405 V3 BLS 50A 30x30 FC&ESC Stack. Flight and Motor controller.
3. 1 x CADDXFPV Ratel2 Analogue Camera. Ratel 2 12000 TVL and 915 MHz receiver Express LRS Nano.
4. 4 x Drone Motors. Emax ecoii series eco ii 2807 6s 1300kv.
5. 1 x Drone antenna. FPV antenna Lollipop 4 V4 SMA 10cm.
6. 4 x Propellers HQPROP 7035 7x3.5x3.
7. Drone Battery. Li-ion Long range pack 6s2p — 9000 mAh.
8. Drone Googles. 5.8G FPV Goggles 40CH X BAND.
9. Radio Module. Bandit Micro ExpressLRS 915MHz RF Module.
10. Radio Controller. RadioMaster TX12 Mark II.
11. Beitian BN-220 GPS module and Buzzer.
12. Payloadfixation and drop system for FPV drones designed to transport payloads weighing up to 3 kg

Software

"Betaflight" open source software, the world's leading multi-rotor flight control software. The global FPV drone racing and freestyle community choose Betaflight for its performance, precision, cutting-edge features, reliability and hardware support. Version, Release Candidate for 4.5.0. www.betaflight.com

12.1.2 Project Build - Step-1

Flight Controller and Electronic Speed Controller

Illustration 1 - SpeedyBee FC & ESC

Flight controller and power distribution board. eq F4 processor, which provides fast and efficient pe controlling the drone's flight and with variety connectors for easy integration with SEC.

Table 1 - Typical features:

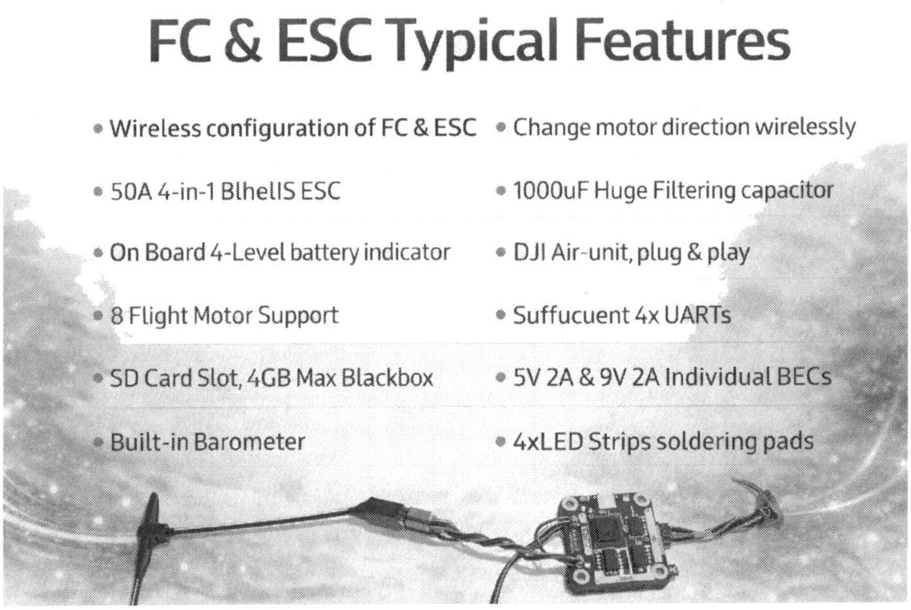

(FC) - Flight controller

(ESC) - Electronic speed controller

12.1.3 Frame Assembly - Step-2

Mark 4 Drone Frame Assembly

Drone Ready-to-Sky Mark 4 - 7" 295 Frame (Assembled)

Illustration 2 - Mark4 V2 Frame

Mark4 V2 7-inch 295mm Quadcopter Carbon Fibre Frame

This style of drone frame is impressively durable and capable of handling repeated crashes without major structural failures. When the carbon-fibre weave is correctly oriented along the arms, it provides excellent stiffness and strength. This alone contributes to months of reliable flight testing—even through several hard landings—offering a strong indication of the frame's overall durability. The increased material thickness further enhances impact resistance and improves mission survivability.

That said, every frame has its limits. During controlled testing, this frame began to show signs of in-flight impact damage at speeds exceeding eighty kilometres per hour when colliding with brick wall targets and steel structures.

Once assembled, the build quality should be carefully inspected to ensure clean, well-machined surfaces throughout. Replacing the soft stock screws with higher-grade steel bolts significantly increases rigidity and results in a more robust and reliable overall construction.

To assemble a Mark 4 drone frame (assuming you're referring to a generic quadcopter frame such as the F450 or similar), here's a step-by-step guide that applies to most Mark 4-style frames:

- Mark 4 drone frame kit (usually includes 4 x arms, top & bottom plates, screws, landing gear)
- Hex driver or small Phillips screwdriver
- Threadlocker (optional, for vibration resistance)
- Motor mounting screws (often M3)
- Power Distribution Board (PDB) or 4-in-1 ESC (if applicable)

Assembly Steps

1. Identify All Parts

Unpack your Mark 4 frame kit and identify:

- 4 x arms (usually plastic or carbon fiber)
- 1 x bottom plate
- 1 x top plate
- Screws and standoffs
- Optional: landing gear, dampers, spacers

2. Attach Arms to Bottom Plate

- Position each arm at the correct corner of the bottom plate.
- Use the provided screws to secure each arm in place.
- Optional: Apply thread locker to screws.

3. Mount the Top Plate

- Place standoffs between the bottom and top plates.
- Screw the top plate into the standoffs, ensuring all arms are sandwiched firmly.

12.1.4 EMAX Motors - Step-3

When the frame is ready, the next step is to prepare and install the motors. use EMAX ECOII Series ECO II 2807 6S 1300KV and a special protective coverage for the motor wires.

Illustration 3 - ECO II Motors & Mounting

All 4 EMAX ECOII motors installed on Mark 4 7" drone frame before soldering. .

EMAX ECOII Series ECO II 2807 6S 1300KV **Motor**

EMAX motors installation using M3 Hex - Motor Mounting

- Mount each motor onto the end of its corresponding arm using the provided screws.
- Ensure motor wires are facing inward toward ESCs/ PDB.

12.1.5 Power Board - Step-4

Illustration 4 - ECO II Motors soldering

All 4 EMAX ECOII motors installed on Mark 4 7" drone frame - before soldering.

Install Power Distribution Board

- Mount PDB in centre of bottom plate.
- Use nylon standoffs or vibration dampers.
- Route wires outward toward ESC.

Illustration 5 - ESC power supply

Soldering the power cables is often the most challenging part of the build. If you're not confident in your soldering skills, take time to practice on scrap wires before attempting this step. Use a high-quality flux-core solder, and set your soldering iron to a temperature between 300–360 °C to ensure strong, reliable joints.

12.1.6 ESC Power Soldering - Step-5

Remember to solder the 1000uF low ESR Capacitor included in the Speedy Bee F405 V3 Stack package. Connect the Red wire to red (+) and the Black wire to black (-). This is necessary to protect the ESC board.

Motor Connections for Mark 4 7" Quad-Copter - EMAX motors soldering

Illustration 6 - ESC power supply connections

Soldering Motor Identification & Connection

(Quad-Copter X-Configuration)

Assuming you're using a standard X-configuration (which is typical for a 7" quad like the Mark 4), here's how the motors are identified and how they should be connected to the flight controller:

Table 2 - Motor connections

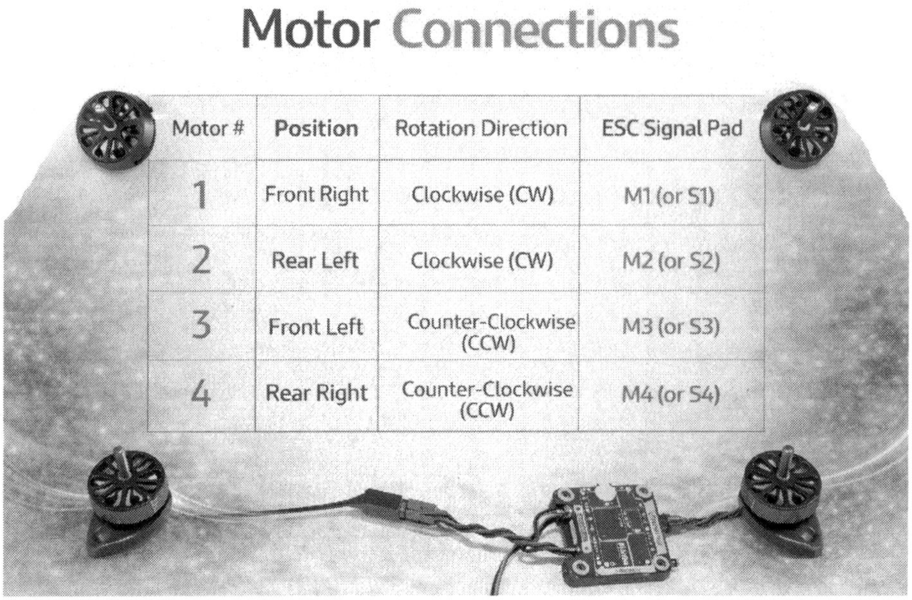

Motor Connections

Motor #	Position	Rotation Direction	ESC Signal Pad
1	Front Right	Clockwise (CW)	M1 (or S1)
2	Rear Left	Clockwise (CW)	M2 (or S2)
3	Front Left	Counter-Clockwise (CCW)	M3 (or S3)
4	Rear Right	Counter-Clockwise (CCW)	M4 (or S4)

Motor Wire Connections - Each motor has three wires:

- Connect them to the three pads on the ESC.
- The order doesn't matter initially—test motor direction after setup. Cut the wires to length after you are sure that the connections are orientated correctly.
- Use your flight controller software (e.g., Betaflight) to check and correct motor direction.
- If a motor spins the wrong way, reverse any two of the three wires.

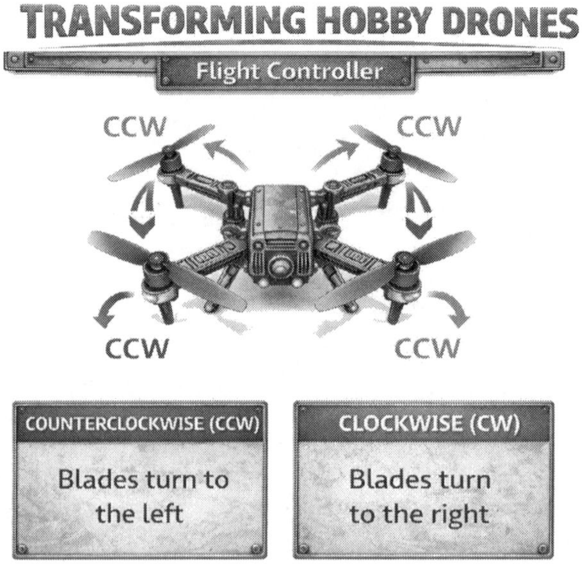

12.1.7 Flight Controller - Step-6

Flight Controller Orientation

- Make sure you know the orientation of the flight controller (arrows usually point forward).
- The motor numbering assumes the FC is installed facing forward with standard orientation.

VTX and camera setup

Reference the Speedy Bee website for the wiring diagram and Annex..A, the Video Transmitter VTX AKK Race Ranger 1.6W should be connected to the 9V, G, VTX, and T1 pins on the first UART port of the FC. T1 means that this is a TX from UART1.

Illustration 7 - VTX Camera connections

Double-check all the pins when connecting any peripheral device to the flight controller because they may vary from device to device. For the VTX, we use red (1) and black (2) wires, as well as yellow (5) and green (6). The internal red (3) and black (4) wires should not be used. I have isolated them. These may be used to connect to the video camera directly, but we will be using the FC to connect with the camera through that.

Video-camera Caddx Ratel 2 12000 TVL connects to the flight controller using 5V, G, and CAM pins as shown in the picture above. Red (1) to 5V, Black (2) to G, and Yellow (3) to CAM.

After soldering all the necessary connections, you can proceed to connect the FPV system either to your computer or directly to the battery for testing. If there is no smoke, congratulations, you have completed the work correctly. From this point, you can even turn on your FPV glasses and search for the channel to search your FPV drone. It should already be operational by this point if the UART1 is configured.

[IMPORTANT]

Do not forget to connect your FPV antenna, Lollipop 4 V4 SMA 10cm, if you power on the VTX without the antenna, there is a chance it could burn out the transmitter circuit.

ELRS receiver Configuration

The ELRS receiver is used to connect your drone with the Radio Controller. The operator manages the drone's direction, speed, turns, altitude, and everything else using this radio controller. It is the primary device for controlling the drone.

For reference, UART stands for Universal Asynchronous Receiver/Transmitter. It is a hardware communication protocol used to transmit and receive data between devices, typically over serial communication. UART is one of the most common ways for micro-controllers, computers, and other devices to exchange data.

How UART Works:

Asynchronous Communication:

- UART is "asynchronous," meaning it doesn't use a clock signal to synchronize the data transmission. Instead, both devices (transmitter and receiver) must agree on certain parameters (like baud rate, start/stop bits, and data bits) to ensure the communication works correctly.

Transmission:

- Data is sent one bit at a time over a single wire (or two wires for full-duplex communication: one for transmitting and one for receiving).

Format:

- Start Bit: Signals the beginning of data transmission.
- Data Bits: The actual data being transmitted (common 8 bits, but it can be configured as 5, 6, 7, or 9 bits).
- Parity Bit: Optional, used for error checking.
- Stop Bit(s): Signals the end of the transmission. There can be one or more stop bits (usually 1 or 2).

Baud Rate:

- This is the speed at which data is transmitted. Both devices need to communicate at the same baud rate, e.g., 9600, 115200 baud, etc.

Full-Duplex:

- UART can operate in full-duplex mode, meaning it can send and receive data simultaneously. However, many implementations only use it in half-duplex mode, where data can be sent or received but not both at the same time.

12.1.8 ELRS Receiver -Step-7

Illustration 8 - ELRS receiver

According to the official documentation, the ExpressLRS (ELRS) receiver should be connected to a designated UART on the flight controller, most commonly UART2, although this can vary depending on the specific FC layout and firmware configuration. It is critical at this stage to carefully double-check which UART you intend to use and confirm that it aligns with both the manufacturer's wiring diagram and your planned configuration in Betaflight. Using the wrong UART or reversing signal lines is a common source of receiver issues and can result in no communication between the receiver and the flight controller.

In this build, an ExpressLRS Nano receiver is used, which simplifies wiring due to its compact form factor but still requires precise connections. The black wire is connected to ground (G) to provide a common reference, the red wire is connected to the regulated 4V5 pad to supply stable power to the receiver, the white wire is connected to R2 (RX on UART2) to carry telemetry data from the receiver to the flight controller, and the yellow wire is connected to T2 (TX on UART2) to transmit control data from the flight controller back to the receiver. Ensuring correct polarity, secure solder joints, and clean routing of these wires will improve signal reliability and reduce the risk of intermittent faults, particularly in high-vibration or long-duration flight scenarios. Once wired, this UART will later be enabled for Serial RX in the Ports tab, allowing the ELRS system to provide its low-latency, long-range control link essential for stable and responsive flight.

CHAPTER 16

DIY Build, Project Software

13.1.1 BETAFlight - Software

Betaflight is one of the most widely used open-source flight-controller firmware platforms for hobby drones. It is known for its high performance and extensive feature set, including on-screen display (OSD), Blackbox logging, and support for a broad range of aircraft configurations. In the context of hobby drones used in warfare, Betaflight has been widely adopted for both modifying existing platforms and building systems from scratch.

As an open-source platform, its accessibility has proven particularly valuable for adapting and extending the base code to support emerging operational requirements, including the integration of artificial intelligence (AI) and machine-learning (ML) capabilities in field deployments. These characteristics have made Betaflight highly configurable across most hobby-style drones employed in conflict environments, including quad-copters, fixed-wing aircraft, and even marine or surface-based unmanned platforms.

[IMPORTANT]

Betaflight — for the latest updates and the web-based configurator setup guide, visit www.betaflight.com. You MUST obtain the latest version for your system, this DIY project is a GUIDE.

Recently, Betaflight transitioned from a traditional version numbering system (for example, 4.x) to a date-based release format (such as 2025.12). This new system provides clearer release identification, with major updates released twice per year in June and December. The change also aligns the firmware and configurator versions, making updates simpler and improving compatibility.

Ongoing development continues to focus on expanding capabilities such as autonomous flight features, improved fixed-wing support, and ongoing user interface enhancements.

Betaflight is an open-source flight controller firmware for multi-rotor drones and fixed-wing aircraft, developed and maintained by the global FPV community. It is free to use and continually improved through community contributions.

Betaflight Configurator – How to Choose the Right Version

[Important]:

The Betaflight Configurator web app only works in Google Chrome or other Chromium-based browsers, such as Microsoft Edge or Vivaldi. It will not work in Firefox, Safari, Opera or any Onion style encrypted browser.

Option 1: Web Version (Recommended for Most Users)

- Open Google Chrome (or another Chromium browser).
- Launch the Betaflight Configurator web app.
- No installation required.
- Always uses the latest version automatically.

This would be the best choice if:

- You want the quickest and easiest setup.
- You have an internet connection.
- You are running a recent version of Betaflight firmware.

Option 2: Desktop Version (Advanced or Offline Use)

If you prefer a traditional desktop application, you can download it here:

https://github.com/betaflight/betaflight-configurator/releases/latest

Choose the desktop version if:

- Your drone is running an older Betaflight firmware and requires a matching Configurator version.
- You don't have internet access, or you need a fully offline setup.
- You prefer a standalone application for field or workshop use.

Tip: If you are unsure which version to use, start with the web version. You can always switch to the desktop version later if needed

Purpose

Betaflight is designed to deliver high-performance, precise, and reliable flight control. Users can customise and tune their aircraft using the Betaflight Configurator, allowing optimisation for different air-frames and flying styles.

Key Features

- On-Screen Display (OSD) for real-time flight information
- Blackbox logging for flight analysis and tuning
- Broad hardware support, including gyros and additional sensors
- Custom flight modes, such as Turtle Mode
- Advanced tuning tools for performance optimisation

Betaflight Versions & Versioning- Previous Versioning System

Earlier Betaflight releases used a number-based versioning scheme (for example, 4.4 or 4.5), with major updates released periodically.

Current Version System

Beginning with Betaflight 2025.12, Betaflight adopted a date-based versioning format (YYYY.MM). Major releases now occur twice per year, in June and December (for example, 2025.12 and 2026.6).

Configurator Alignment

The Betaflight Configurator follows the same date-based versioning as the firmware. This alignment simplifies updates and helps ensure compatibility between firmware and configurator versions.

Development Cycle

New features progress through the following stages before official release:

- Alpha – Active development and experimentation
- Beta – Feature-complete with a focus on stability
- Release Candidate (RC) – Final testing before public release

Recent (2025.12) & Upcoming Features

- Autonomous & Safety Enhancements
- Altitude and Position Hold
- Collision detection
- Automatic disarm for improved safety

Fixed-Wing Improvements

- Throttle and PID attenuation
- S-term support
- Airspeed-based flight modes

Flight Experience Improvements

- Enhanced Turtle Mode behavior
- New and improved timers
- LED dimming support
- Compatibility with new gyros and sensors

Betaflight Configurator Updates

- Progressive Web App (PWA) for automatic updates
- Web-based CLI access
- Improved DFU mode support

13.1.2 BETAFlight - How to Set Up

Betaflight configuration involves installing the software, connecting your craft, and systematically going through the various tabs in the Configuration to set up the hardware and flight parameters. The process typically requires both physical wiring and software settings. Before configuring your craft, ensure you have the required software installed. You will need the Betaflight Configurator for your operating system (Windows, macOS, or Linux). This application is used to configure and manage the settings on your flight controller.

Download and install the Betaflight Configurator before proceeding.

Connecting to Your Flight Controller:

1. Open the Betaflight Configurator.
2. Connect your flight controller to your computer using a USB cable.
3. In the top-right corner of the configurator, select the appropriate COM port from the dropdown menu.
4. Click Connect.

Once connected successfully, the configurator will display the main dashboard, showing system information and sensor status.

Illustration 9 - BETAFlight - Home Page

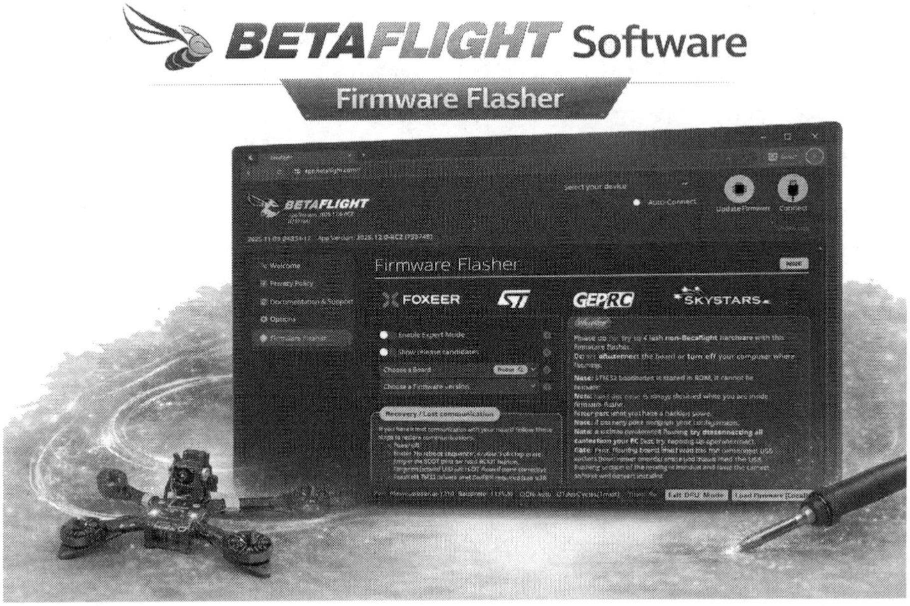

NOTE: If no COM port appears, ensure the correct USB drivers are installed and try a different USB cable or port.

Illustration 10 - BETAFlight - Com Port Connection

[Example] When you have a COM port selected, click the "Connect" button.

When connected the "connection" button should turn green.

If you do not see a new COM port appear, or the configurator cannot connect, there are a few ways to solve it:

1. Make sure that you are plugging the USB cable into the flight controller, nothing else. Do not connect the Betaflight App to an HD system. Do not connect the Betaflight App to a radio transmitter. the Betaflight App is not meant be used with anything other than a flight controller
2. Make sure you are using a USB cable that is capable of data transfer. Some USB cables are only for charging
3. You may need to install the drivers for your flight controller. There is a download link for the ImpulseRC Driver Fixer tool in the configurator, or you can download it from here
4. If you are still experiencing issues, try shutting down/uninstalling any other software that may be using the COM port. 3d printing software is a common culprit

13.1.3 BETAFlight - Configurator Options

Before connecting your flight controller, let's adjust a few initial Configurator settings to make your workflow smoother.

Click Options in the left-hand panel.

Here are a few recommendations:

1. Reopen last tab on connect - automatically opens the last tab you used whenever you reconnect to the flight controller.
2. Virtual connection mode - lets you explore and learn Betaflight Configurator without needing a real flight controller connected.
3. Enable Dark Theme - easier on the eyes, especially for long setup sessions.
4. Change language - switch to your preferred language for better readability.

Illustration 11 - BETAFlight - Options

In summary, the Main Options tab provides access to a short, streamlined list of essential configuration settings that control the most important aspects of the flight controller. This tab is designed to give pilots quick access to core options without navigating through advanced or specialist menus. It typically includes fundamental system settings that affect how the flight controller operates, such as enabling key features, selecting basic behaviours, and confirming that the aircraft is correctly configured at a high level. By keeping this list concise, the Main Options tab allows users to verify critical settings quickly, making it especially useful during initial setup, troubleshooting, or pre-flight checks.

13.1.4 BETAFlight - Updating Firmware

Do You Need to Update Betaflight Firmware?

You don't need to update the firmware right away.

If you purchased your flight controller using my recommendations, it should already have Betaflight pre-installed. Even if a newer version is available, updating is optional, not required:.

1. When not to update
2. Your drone is flying well.
3. You are new to Betaflight.

You don't need any new features or fixes from the latest release.

[Important]:

Updating (flashing) Betaflight firmware will usually erase all your current settings. This means you'll need to reconfigure the entire drone from scratch, including ports, receiver, modes, and tuning:

1. When it does make sense to update
2. The new version includes a specific feature you want.
3. It fixes a bug that affects your setup.
4. You are comfortable reconfiguring the drone afterward.

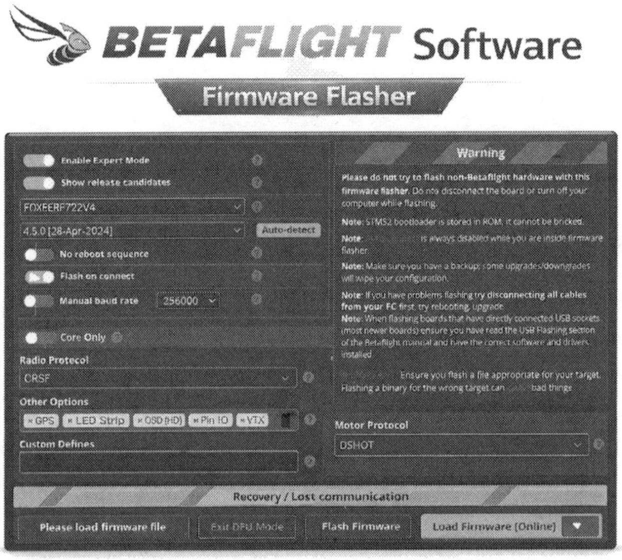

If you decide to update

Follow this step-by-step guide before flashing anything:

https://betaflight.com/docs/wiki/guides/current/Installing-Betaflight

Tip: If you're unsure, leave the firmware as it is. A stable, flying drone is better than a newer version that isn't set up yet.

13.1.5 BETAFlight - Connecting Your Drone

Connect your drone to your computer using a USB cable — make sure you're plugging into the actual flight controller, not the DJI Air Unit or other components. There is no need to plug in the battery.

If your computer doesn't recognise the flight controller, double-check that you're using a data cable, not a charge-only cable.

In Betaflight Configurator:

1. Click "Select your device", then "I can't find my USB device".
2. A pop-up will appear showing available COM ports.
3. Select the one that says "Betaflight".

Tip: If you're unsure which COM port is the right one, unplug the USB cable and reconnect it — the correct port is the one that appears when you reconnect. If you can't find the COM port or still can't connect, you may have a driver issue on your computer. Follow this Flashlight trouble shooting guide : https://www.betaflight.com/docs/wiki/getting-started/troubleshooting.

13.1.6 BETAFlight - Interface Explained

Once you click Connect, the Setup tab will open and display a 3D model of your drone. This model moves in real time as you tilt or rotate the flight controller, allowing you to confirm that the orientation and sensors are working correctly.

Illustration 15 - BETAFlight - Interface Explained

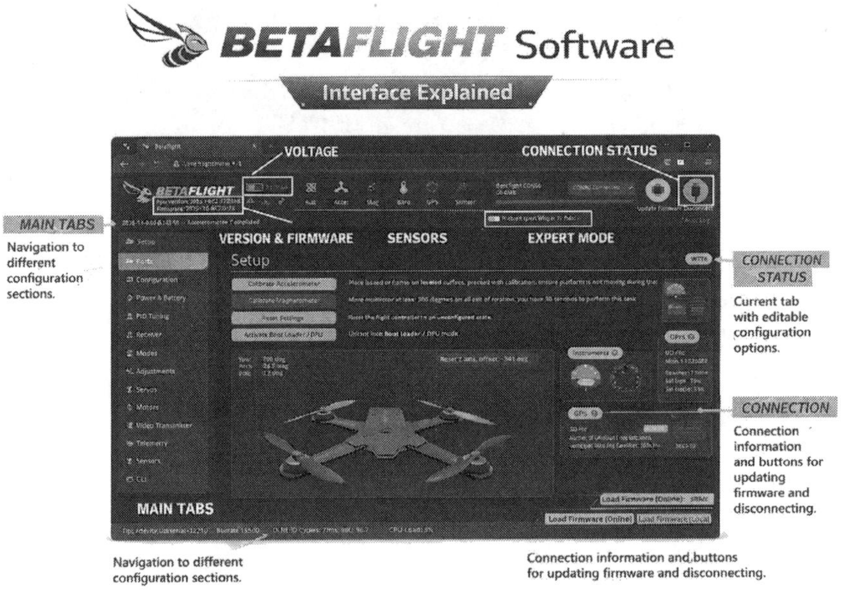

Taking a quick tour of the interface and focusing on the key items you should pay attention to.

Top Section (Status Information)

- Firmware Version & Target. Shows the Betaflight version installed and the target flight-controller board (FC) you're using.
- Voltage Reading
  Displays the voltage detected by the flight controller.
- If no LiPo battery is connected, this will usually show the USB voltage instead.
- Sensor Icons (Top Bar)
  The icons indicate which sensors on your flight controller are active.
- Lit icons = sensor detected and working
- Red icon = sensor issue or not ready
- Example: A red GPS icon usually means it hasn't locked onto satellites yet.

[Note] For basic flying, only the Gyro is required.

Other sensors enable extra features or flight modes but are not essential to get airborne.

Left-Hand Menu Items

On the left side, you'll see a list of configuration tabs. To keep things simple, we'll only work with the essential tabs needed to get your drone flying. Once you're more comfortable with Betaflight, you can explore the additional tabs to unlock advanced features.

Enable Expert Mode

Before going any further, click Enable Expert Mode in the top-right corner.

- This reveals additional tabs and advanced settings that are hidden by default.
- You don't need to change these settings right now.
- Enabling Expert Mode prevents confusion later when you're following guides or looking for settings that aren't visible.

Common Warning Message

You may see a popup warning such as:

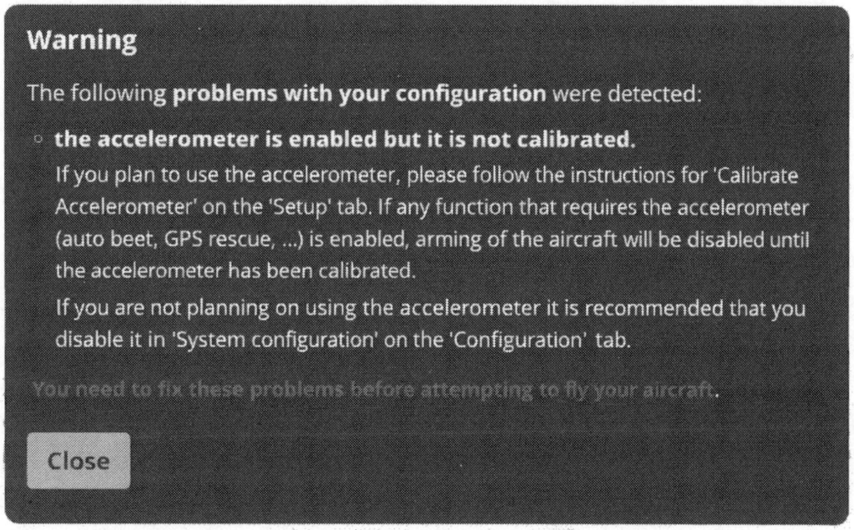

"The accelerometer is enabled but not calibrated." Don't worry — this is normal. To fix it:

- Place your drone on a flat, level surface.
- Go to the Setup tab.
- Click Calibrate Accelerometer.
- Once calibration is complete, the warning will disappear.

Betaflight Configurator – Initial Setup Checklist

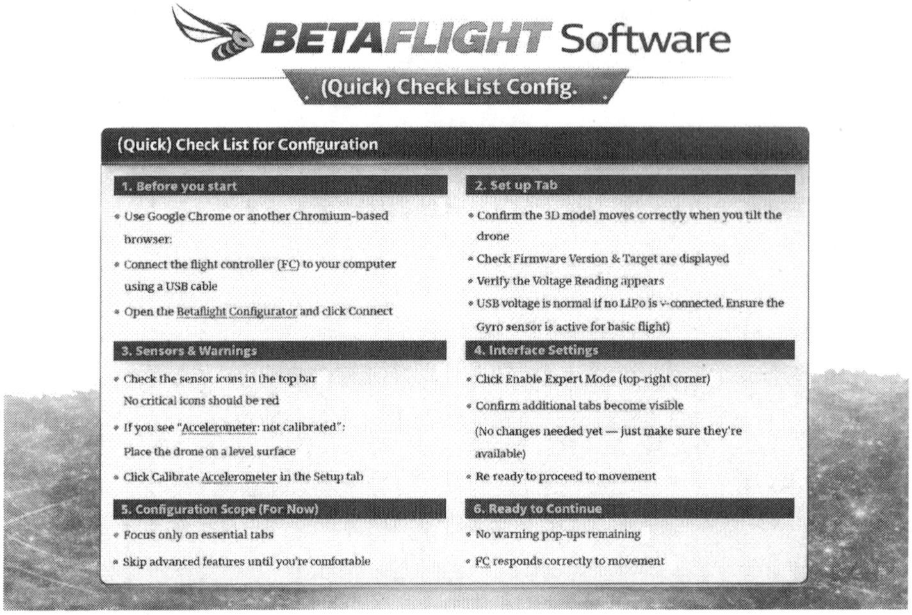

13.1.7 BETAFlight - Preset Tabs

[Important]Before making any changes to your flight controller, it is essential to back up the current configuration. This ensures you can recover from mistakes or revert to known-good settings if required.

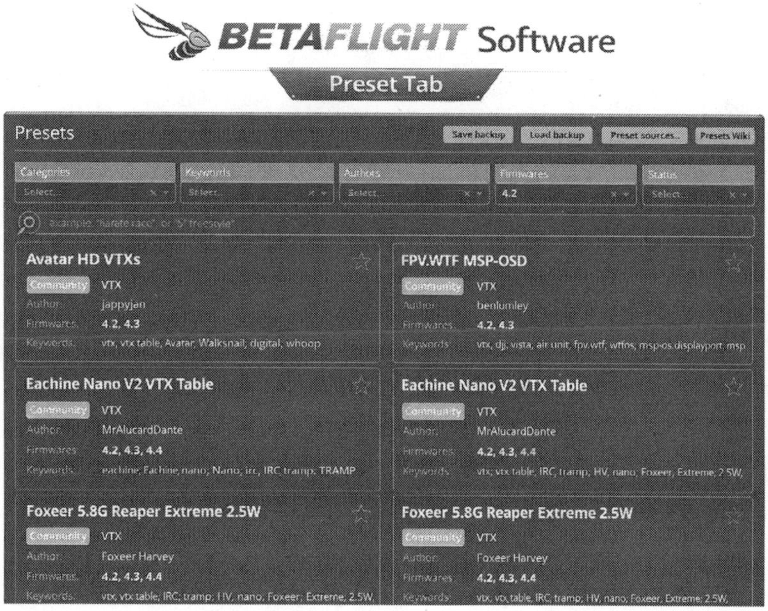

Betaflight configurations can be backed up in several ways (see: https://www.betaflight.com/docs/wiki/getting-started/setup-guidebook-your-config).

The easiest and most reliable method is to use the Presets tab.

- Connect your flight controller to Betaflight Configurator.
- Navigate to the Presets tab.
- Click "Save Backup" to create a complete backup of your current configuration.

This backup stores all flight controller settings and can be restored later if the firmware is updated, settings are corrupted, or a rollback is required.

The Presets tab also provides access to community-shared presets, including:

- Flight tunes
- Rate profiles
- OSD layouts

These presets can be browsed and applied directly within Betaflight Configurator without manually entering CLI commands. To apply a preset, select it from the list, review the included changes, and apply it to your flight controller.

13.1.8 BETAFlight - Setup Tabs

The Setup tab is the first screen displayed when you open Betaflight Configurator. It is used to verify basic sensor operation and flight controller orientation.

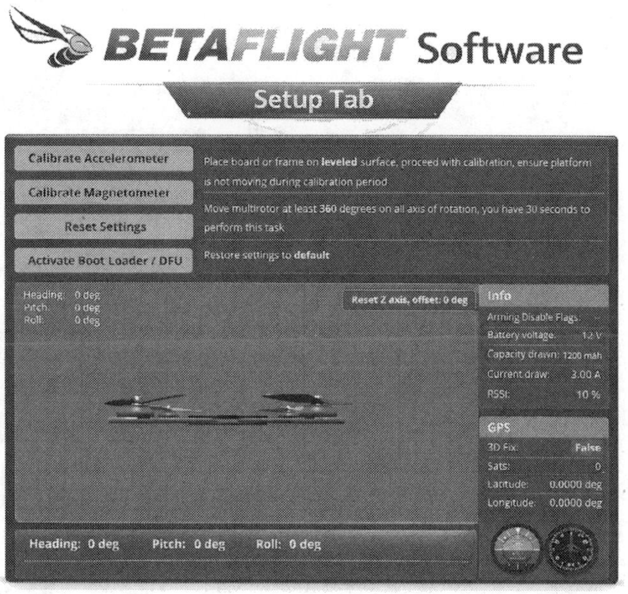

To check the accelerometer and board alignment:

- Connect your flight controller to Betaflight Configurator.
- Open the Setup tab.
- Hold the drone securely in your hand with the camera facing away from you.
- Click "Reset Z Axis" to level the accelerometer.
- Slowly tilt and rotate the drone in all directions.

The 3D model on the screen should move in the same direction and orientation as the physical drone. If the on-screen model matches your movements, the accelerometer is functioning correctly and the flight controller orientation is set properly. If the model moves in the wrong direction or appears misaligned, the board orientation must be corrected before flying, as incorrect alignment can cause unstable or uncontrolled flight.

13.1.9 BETAFlight - Ports Tab

You will have multiple different devices and peripherals connected to your flight controller UART ports, like a GPS, VTX control, or even a wireless adapter. This is where you tell the flight controller how to read, and what to do with the data it receives from these ports

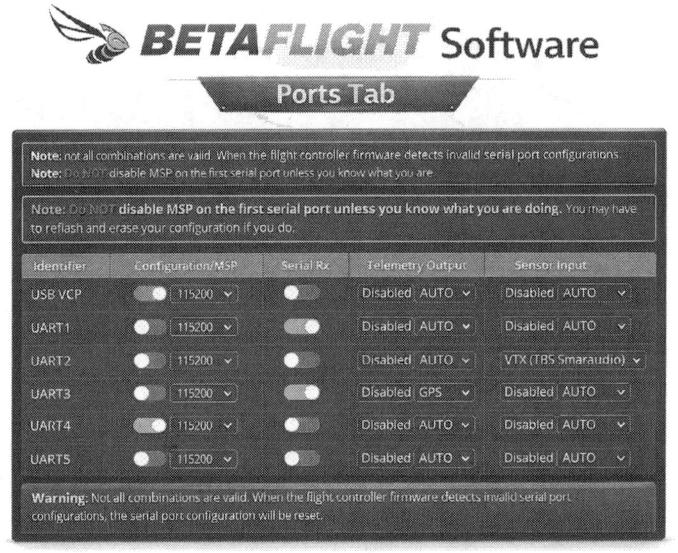

The Identifier is simply the label for each communication port on the flight controller (FC).

You will usually see labels such as:

- UART1, UART2, UART3, etc.
- SOFTSERIAL
- USB / VCP

Each UART number corresponds to a physical RX-TX pin pair on the flight controller. When you connect a device (GPS, receiver, VTX, telemetry module), it must be wired to the matching RX/TX pins for that UART and configured here.

Key points:

- One UART can support multiple functions, but only one serial device per RX/TX pair
- USB/VCP is used for communication with the Betaflight Configurator
- SOFTSERIAL emulates a UART in software (lower performance, used when hardware UARTs are exhausted)

Configuration / MSP

The Configuration / MSP column controls whether a port is used for MSP (MultiWii Serial Protocol) communication.

MSP is a low-level communication protocol that allows:

- Configuration via Betaflight Configurator
- Control and telemetry from external devices
- Direct interaction with the flight controller by companion computers or peripherals

Typical MSP uses include:

- USB connection to Betaflight Configurator
- On-screen display (OSD) devices
- Companion computers (e.g., telemetry relays, mission controllers)
- External configuration tools

You can also manually set the baud rate, which defines how fast data is transmitted:

- Higher baud rates = faster communication
- Must match the connected device's supported baud rate
- Incorrect baud rates cause connection failures or unstable data

Important rules:

- MSP should only be enabled where needed
- Enabling MSP on unnecessary ports can cause conflicts
- USB/VCP almost always has MSP enabled by default

BETAFlightt Port Tab

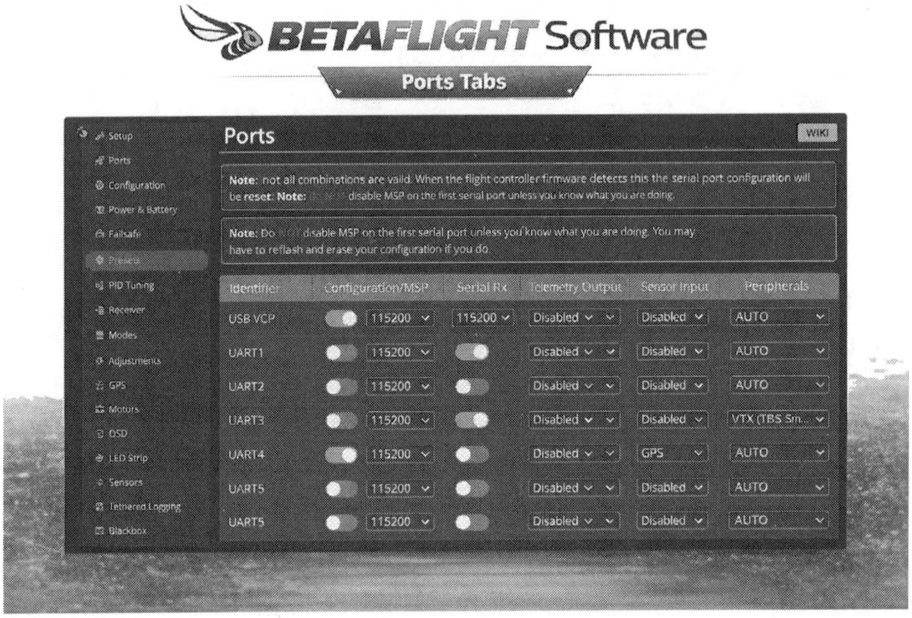

13.1.10 BETAFlight - Configuration Tabs

This is where the majority of your flight controller's core configuration takes place.

Here you will set critical parameters such as PID loop timing and gyro update frequencies, enable or disable onboard sensors not configured in the Ports tab, and adjust a wide range of other fundamental system settings that directly affect flight performance, stability, and reliability.

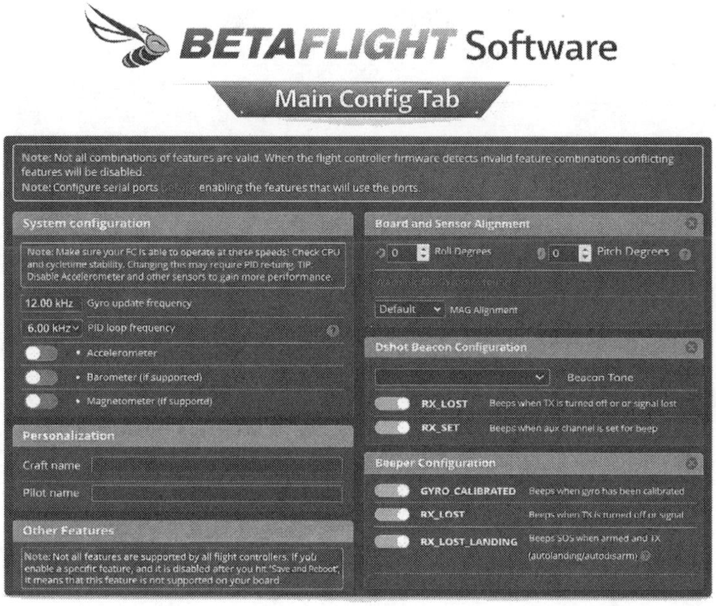

System Configuration

Gyro Update Frequency

The rate at which the gyro is sampled. In most recent versions of Betaflight, it will default and lock to the frequency that the gyro runs best at (8KHz for the MPU6000, 3.2KHz for the BMI270, etc...)

PID Loop Frequency

The frequency at which the PID loop computations are done. This is basically all of the math that goes into actual flight control. When using dshot300 you may see the PID loop reset to 4K if you manually attempt to set 8K, dshot300 does not send updates fast enough to make use of 8K PID loop so 4K is selected to save CPU time.

Recommended PID loop and motor output combinations, With RPM filtering enabled, are 2K/dshot150, 4K/dshot300 and 8K/dshot600. Exception is when using the BMI270 gyro in which case the rates are 3.2K/1.6K

[INFORMATION]

Usually it's best to have it set to the same frequency as the gyro, or a half of it if you are using a slower MCU and a high gyro frequency (8KHz gyro would be 4KHz PID loop on an F411)

Sensor Toggle

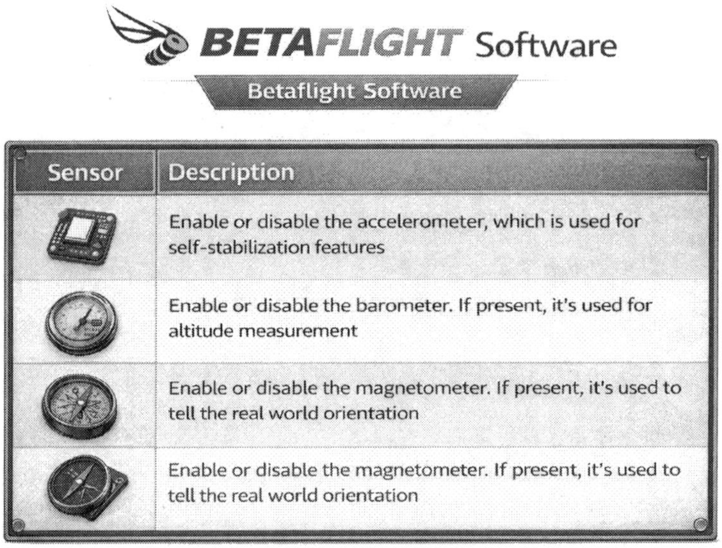

Accelerator Trim

Trim the accelerometer to compensate for any errors in the accelerometer readings. Only visible when the accelerometer option is enabled.

Camera

Used to set the camera angle to be used for things like FPV angle mix mode and artificial horizon. Only visible when the accelerometer option is enable.

Arming

Set a maximum angle that the craft can be tilted at while arming, to prevent arming the craft when it's in an unsafe position

[INFORMATION]

Unless you specifically need a fail-safe like this, setting it to 180 degrees will disable the check, and you can arm the craft in any position, useful for flip over after crash

Board and Sensor Alignment

Allows you to virtually offset the FC and other sensors if they're mounted in a non-standard way

[TOP-TIP]

If the 3d model preview is not responding correctly to real world movement, it's most likely because the board alignment is incorrect. Use the alignment options to fix it. Go in increments of 90 degrees (45 if it's a diagonally mounted FC) and test the preview after each change

Personalisation

Allows you to set a craft and pilot name to be shown in the OSD, blackbox logs and diff/dump outputs

DShot Beacon Configuration

Running a high frequency signal on the motor output to make the motors resonate and make a sound. A good alternative for an actual buzzer, but it cannot be activated in flight (as the motors are spinning), isn't as loud, and can't be used for a long time as the motors may draw excessive current and overheat

BETAFLIGHT Software

Betaflight Software

Setting	Description
Beacon Tone	You can pick from 5 different tones
RX_LOST	Enable or disable the tone when the RX signal is lost
RX_SET	Enable or disable the tone when the BEEPER mode is on
RX_SET	Enable or disable the tone when the BEEPER mode is on

Other Features

A list of various different features that can be enabled or disabled, and may or may not be present on your flight controller

BETAFLIGHT Software

Betaflight Software

Setting	Description
AIRMODE	Permanently enable airmode, which will give the craft more control authority in the air when at 0 throttle
CHANNEL_ FORWARDING	Allows you to forward an aux channel to a motor/servo output
DISPLAY	Enables the display feature, which allows you to use a small OLED display to show various information. If enabled and no display is connected, the FC will take about 10s longer to boot. It's not recommended to use nowadays
GPS	Enable GPS support, do not enable when there is no GNSS unit attached
INFLIGHT_ACC_CAL	Allows you to calibrate the accelerometer in flight
LED_STRIP	Enables the LED strip feature, which allows you to control WS2812B RGB LEDs
OSD	Enables the OSD, you can configure it in the OSD tab that will appear when you enable this
SERVO_TILT	Enables the CAMSTAB mode, which will stabilize the camera angle up by to two servos set in a gimbal configuration
SOFTSERIAL	Emulates a serial port on a different output, or splits an RX-TX UART pair. Allows you to use it as an extra UART
SONAR	Enables sonar support, but this feature is not recommended for use nowadays
TRANSPONDER	Enables the race transponder feature if your hardware supports it

[INFORMATION]

Softserial is useful for FCs that don't have enough UARTs to support all of the features you want to use. However, there are some limitations.

- It runs at a lower baud rate. Works well at 9600, but not so well on higher rates
- Consumes more CPU resources, and puts an extra load on the CPU. So not ideal for lower performance MCUs
- It's not ideal to run a receiver on a soft-serial port due to duty cycle limitations
- Some ports may work better than others. LED_STRIP usually works all the time, but you may need to experiment with others
- And lastly, you cannot have more than two soft-serial ports active at the same time

Beeper Configuration

Toggle different triggers when the beeper should be active

BETAFLIGHT Software

Betaflight Software

Setting	Description
GYRO_CALIBRATED	Beeps when the gyroscope has been calibrated
RX_LOST	Beeps when the RX signal is lost (repeats until the signal is regained)
RX_LOST_LANDING	Beeps when the RX signal is lost, and the craft is in a landing phase
DISARMING	Beeps when the craft is disarmed
ARMING	Beeps when the craft is armed
ARMING_GPS_FIX	Beeps when the craft is armed, and the GPS has a fix
BAT_CRIT_LOW	Beeps when the battery is critically low according to the value set in the Battery tab (repeats)
BAT_LOW	Beeps when the battery is low according to the value set in the Battery tab (repeats)
GPS_STATUS	Beeps x amount of times depending on the amount of satellites found
RX_SET	Beeps when the BEEPER mode is on
ACC_CALIBRATION	Beeps when the inflight accelerometer calibration is successful
ACC_CALIBRATION_FAIL	Beeps when the inflight accelerometer calibration fails
READY_BEEP	Beeps when the craft has a GPS fix and is ready to be armed
DISARM_REPEAT	Beeps when the sticks are in a disarm position
ARMED	Beeps when the craft is armed and motors are not sinning (repeats until throttle is raised, or the craft is disarmed)
SYSTEM_INIT	Beeps when the FC has been powered on

13.1.11 BETAFlight - Voltage Calibration

BETAFlight, the Power and Battery tab is used to configure and monitor all settings related to electrical power, battery management, and energy consumption. This tab includes options for setting the battery type and cell count, either manually or via auto-detection, ensuring voltage readings and warnings are accurate. It allows configuration of voltage thresholds for minimum, warning, and maximum levels, which are used to trigger OSD alerts and failsafe behaviour. The tab also includes current sensor calibration and capacity-based monitoring, enabling accurate measurement of current draw and milliamp-hour (mAh) consumption for safer battery management.

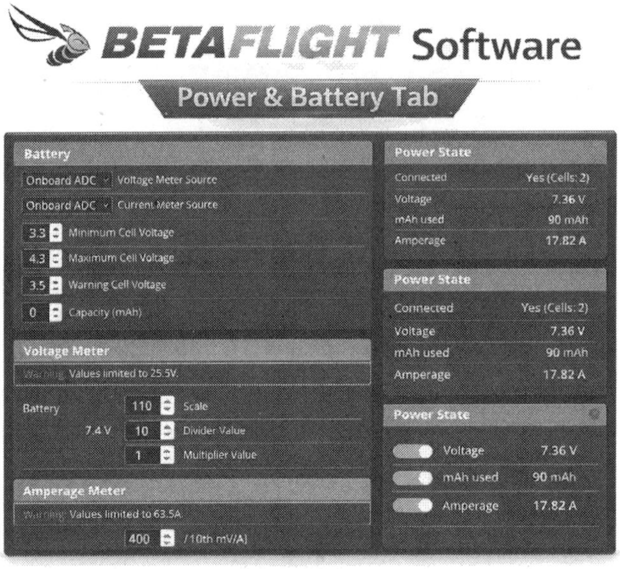

Additional settings include VBAT sag compensation to counter voltage drop under load, power scaling and filtering to stabilise readings, and support for power sources such as external BECs or ESC telemetry. Together, these settings help pilots protect batteries from over-discharge, maintain consistent flight performance as voltage drops, and gain clear insight into real-time and post-flight power usage, making the Power and Battery tab essential for both flight safety and system reliability.

13.1.12 BETAFlight - Failsafe Tab

Fail safe

Fail-safe is the action your craft takes when it loses connection to the transmitter. This tab allows you to configure what your craft does in an event of a fail-safe.

[DANGER}

It is very important to set up your fail-safe correctly to prevent your craft from flying away or crashing when you lose connection.

Fail- safe Stages

- Stage 1 - entered when a flight channel has an invalid pulse length, the receiver reports fail-safe mode or there is no signal from the receiver for 150ms, the channel fallback settings are applied. If valid data is received, Stage 1 fail-safe will terminate automatically. Stage 1 duration is configured using the failsafe_delay duration, and defaults to 1.5s.

- Stage 2 - entered when radio link is lost for more than the configured Stage 1 duration, and the quad is armed. Stage 2 Fail-safe Mode will be activated. Channels that are not controlled by the fail-safe mode remain at their fallback setting.

[Note]

Prior to entering stage 1, channel fallback settings are also applied to individual AUX channels that have invalid pulses.

Valid Pulse Range Settings

Minimum length - The minimum length of the valid pulse range. If a channel goes below this value, the FC will be put into fail-safe mode.

Maximum length - The maximum length of the valid pulse range. If a channel goes above this value, the FC will be put into fail-safe mode. For most cases, leave these settings at their default values.

Fail-safe Switch

This setting determines which stage of fail-safe will be activated when manually engaging the Fail-safe mode from your controller using the Fail-safe Aux Mode.

Stage 1 - activates both Stage 1 and then Stage 2 fail-safe, in their normal sequence. This is useful if you want to simulate the exact signal loss fail-safe behavior.

Stage 2 - skips Stage 1 and activates the Stage 2 procedure immediately

Kill - disarms instantly (your craft will crash).

Stage 1 - Channel Fallback Settings

These settings are applied to invalid individual AUX channels or to all channels when entering stage 1.

[NOTE]

Values are saved in steps of 25usec, so small changes disappear

You can set the channel fallback settings for each channel individually.

The available options for each channel are:

- Auto - Roll, Pitch and Yaw will be set to the center position, Throttle will be set to the minimum value (the quad will drop, fast). This is only available for the first 4 channels
- Hold - The channel will hold its last valid value.
- Set - You can set a custom value for the channel.

Stage 2 - Fail-safe Procedure

There are three options for the fail-safe procedure:

- Drop (default) - The craft will disarm and fall out of the sky.

- Land - The craft will attempt to land.
- GPS Rescue - The craft will attempt to return to the home position.

Land options

- Throttle value used while landing - The throttle value the craft will use while landing, unless the firmware includes Altitude Hold, in which case the throttle will be automatically adjusted to provide a controlled descent.
- Delay for turning off the motors during Fail-safe [seconds] - Time to stay in landing mode until the motors turn off.
-

[WARNING]

If Altitude Hold is not included in the firmware:

- the quad will not disarm automatically, and will only disarm after the Landing Delay period has expired, which by default is 60s;
- If the Fail-safe Landing Throttle value is too high, the craft will ascend instead of landing;
- If the Fail-safe Landing Throttle value is too low, the craft will descend rapidly and may crash badly;

GPS Rescue Options

GPS options in a table format, formatted in one row.

BETAFLIGHT Software

Betaflight Software

Option	Description
Altitude mode	• **Maximum Altitude** – The craft will ascend to the maximum altitude during that flight plus GPS_RESCUE_INITAL_CLIMB. • **Fixed Altitude** – The craft will ascend to the configured Return Altitude. • **Current Altitude** – The craft will fly back at the altitude it was at when the failsafe occurred, plus the Initial Climb altitude, or at the Initial Climb Altitude if the failsafe occurred below the Home Altitude.
Initial climb (meters)	The distance the quad wil Vill climb, above the current altitude, when a rescue is initisted and the altitude mode is set to Current Altitude; also added when in Maximum Altitude mode.
Return altitude (meters)	The altitude the craft will fly back at when the altitude mode is set to Fixed Altitude.
Ascent rate (meter/second)	The rate at which the craft will ascend to the necessary altitude.
Return ground speed	The speed at which the craft will return to the home position.
Maximum pitch angle	The maximum pitch angle the craft will use to return to the home position. Rescue throttle usually needs to be increased if the pitch angle is increased. if pitch angle is too low, the craft might not be able to return in high winds.
Throttle minimum	The rate at which the craft will descend to the ground when it reaches the home position.
Throttle maximum	The maximum throttle value the craft will use to return to the home position.
Throttle hover	Make sure that the craft has enough Throttle Maximum to be capable of achieving the set return velocity, and maintaining altitude at high pitch angles, into any expected headwind!
Minimum distance to home (meters)	The initial throttle value that is applied when the rescue first starts. It is important to set this value accurately, if the quad drops every time the rescue starts, this value is probably too low.
Sanity checks	On (highly recommended): The craft will perform sanity checks before starting the rescue; if the checks fail, the craft will disarm.

Testing Fail Safe

[DANGER]

It is very important to make sure your fail-safe works as expected

Failure to do so can result in lost or damaged equipment, or even injury to yourself or others.

- When using the Drop fail-safe, the easiest way to test it is to arm your craft with your props off, then turn off your transmitter. The motors should stop after the configured fail-safe delay time.
- When using other fail-safe procedures, you can test them by settings up the fail-safe mode on a switch, and carefully activating it in an open area. If something goes wrong, you can revert the position of the switch to stop the fail-safe procedure.
- For GPS Rescue, there is a GPS Rescue mode that can be activated independently of the fail-safe mode. This mode will immediately activate Stage 2 GPS Rescue, bypassing Stage 1.

[TIP]

If a switch-induced Landing or Rescue test has a problem, or seems to float around, DO NOT DISARM IN PANIC! Undo the Rescue switch promptly and you'll get full control back immediately. Do not wait more than 15s or it may disarm and crash.

13.1.13 BETAFlight - Receiver Tab

The Receiver tab is divided into two main sections:

Receiver Output Preview

This section provides a real-time visual display of the input signals coming from the receiver. As you move the sticks or switches on your radio transmitter, the corresponding channels should move in the preview. This allows you to verify correct channel mapping, stick direction, centring, and full range of motion, and to quickly identify issues such as reversed channels or missing inputs.

Receiver Configuration

This section is where you define how the flight controller communicates with the receiver. It includes selecting the receiver type and protocol, configuring channel mapping, and enabling the appropriate serial interface if required. Correct configuration here ensures that the FC can reliably decode the receiver signal and translate pilot inputs into accurate flight control.

Together, these two sections ensure that pilot commands are correctly received, interpreted, and passed to the flight control system—making the Receiver tab a critical part of any BETAFlight setup process.

Output Preview

The receiver output preview shows the current state of the receiver channels, both in channel value and graph forms, and how those affect the drone movement.

Receiver Configuration

Used to configure all of the settings that are specific to your receiver

Receiver

Select the communication protocol used by your receiver. The options are:

- PPM/CPPM - Legacy protocol, unlikely to be used in modern setups
- Serial-based - Most modern receivers communicate over serial, using different protocols like CRSF or SBUS
- PWM - Legacy protocol, unlikely to be used in modern setups
- MSP - An advanced option, using the MSP protocol to communicate with the receiver
- SPI - Used for most integrated receivers, like ExpressLRS on tinywhoop AIO boards

[CAUTION]

Selecting the incorrect protocol will lead to no signal being detected, or the signal being interpreted incorrectly. You have to pick the correct one for your receiver

Telemetry

Toggle the telemetry output on or off. Also required for VTX control from ELRS receivers

RSSI

Mostly a legacy option, used to configure a separate analogue 0-3.3V RSSI input. Most modern receivers communicate RSSI (along with other telemetry data) over the same serial connection as the control data.

Do not enable this option with a modern receiver

Channel Map

Different receivers output the four main control channels:

- Aileron - Roll (left/right)
- Elevator - Pitch (forward/backward)
- Throttle - Throttle (up/down)
- Rudder - Yaw (left/right)

[CAUTION]

If your radio input does not match up to what you see in the preview, you need to change the channel map. There are also preset options for some of the more common systems:

- FrSky/Futaba/Hitec - FrSky, Futaba, and Hitec receivers output the channels in the same order as the Betaflight default (AETR1234)
- Spektrum/Graupned/JR - Spektrum receivers output the channels in a different order than the Betaflight default (TAER1234)

RSSI Channel

Some older receivers only had RSSI output on a single channel. If you have an older receiver, you can set which channel is used to read the RSSI value. This is usually AUX 4 or 12. Leave this setting disabled if you have a modern receiver such one using the CRSF or GHST protocols

One use case for this setting on modern equipment is to view LQ on DJI FPV goggles, which do not include a native LQ field. In this case you can set this channel to the channel used for LQ - AUX11 on ELRS. A better solution is to enable the osd_craftname_msgs CLI option or install WTFOS on your DJI FPV system for a full customisable OSD

"Stick" Settings

Minimum/Center/Maximum values for the four main control channels. These are used to set the range of the stick values, usually for safety and calibration purposes

Dead-band Settings

Dead-band is the range of stick movement that is ignored. Some radios/receivers may have a small amount of jitter, and this setting can be used to ignore that. You also have options to set it specifically for Yaw and 3d mode throttle

RC Smoothing

Toggle the RC smoothing filter on or off.

13.1.14 BETAFlight - Modes Tab

Modes are used to enable or disable features and trigger FC actions using AUX channels switches. Modes are enabled when Ranges or Links are active.

Option	Description
ARM	Enables motor output and allows the craft to fly.
ANGLE	Flight mode that remains level using the accelerometer; stick input affects the angle of the craft.
HORIZON	Flight mode that remains level using the accelerometer; stick input affects angle, but at extremes, the craft will flip upside down and back to level.
ANTI GRAVITY	Increases P and I terms during fast throttle movements to improve stick tracking and avoid nose drift.
MAG	Activates heading lock via the magnetometer (compass).
HEADFREE	Flight mode where yaw is aligned with an external reference (often where the pilot is facing); designed for beginners but rarely used – advise ANGLE mode.
HEADADJ	Sets a new yaw orign for HEADFREE mode.
CAMSTAB	Engages servosit to respond to the crafts movements and auto level a gimbal for camera stabil iration,
PASSTHRU	Activates heading lock via the magnetometer (compass).
BEEPER	Skips PID loop and passes roll, yaw, and pitch directly to servos vos for ariplane 'arges.
LEDLOW	Turns-LED sinp off Calibrates rol/pifich offeets for the craft to fly.

Option	Description
OSD	Enables/disables the OSD everlay taause.
TELEMETRY	Enables/disables the sanding of /C relumetny to the control fmt.
SERVO1	Enables/disables the fitis servo output.
SERVO2	Enables/disables the aecond servo output.
SERVO3	Enables/disables the fhint servo output.
BLACKBOX	Enables/disables blackbos log sntermehs, ceahes oot tlo f8e majred date nheat acamghis is himben)
FAILSAFE	Repieates a control his fature event to adomise ofmnoic heating of GTS Reaum 1O inane.
AIRMODE	Enables/disables AF mode whtchae fUll PID convento uss,ke10 sirable Moremeon
3D	Enables recesable motor direction for magines. hrrtirS, oloming.
FPV ANGLE MIX	Heatre0 flight shrode/songe Receurse -1ah lez. 100. Enarncoos to fov. 100.
BLACKBOX_ERASE	Cleare at dias from the bancoes. Rommacoto moirne pradte?
CAMERA CONTROL 1	Cubton action to configure a Rancere. Conainatie curnees Baam rugia innn.?
CAMERA CONTROL 2	Cubton action to contigure a Rancen. Conaitratie curnees baam rugis innnvz?
FLIP OVER AFTER CRASH	Barse festubarher of imoots ARM satbrnis, wonel, oloug stoungd)
	Enables MSP Overrile mode.
MSP OVERRIDE	Enables NSP Overrale mode.
STICK COMMANDS DISABE	Beeper Mute.

Option	
OSD	Enables/disables the OSD overlay tcature.
TELEMETRY	Enables/disables the ecating of tekeevery to the control inris ioeeven.
SERVO1	Enables/disables the servo output.
SERVO2	Enables/disables the serrvo output.
SERVO3	Enables/disables the thint servo output.
BLACKBOX	Enables/disables bisxtoes log recording haith to dowit to camens pecontertalee.
3D	Enables rerosible morje direction for magnag mace, alowing memas ragis mode.
FPV ANGLE MIX	higremetre that apoirec you cowition fesative oe tis waelable.
BLACKBOX ERASE	Egers al flits from the blackoos, thishmorozo be algy alfated.
USER1	User-detnool awitth 1., comnots arb itory output ust PMAX.
USER2	User-detnool swith 2., comnots arb itory output ust PMAX.
USER3	User-detnool awith 3., comnots arbitory output ust PMAX.
PID AUDIO	Enables output of PIUs onrroler suse as audio
PARALYZE	Permonerfy disables a downed craft unel it is cower nocoro.
GPS RESCUE	Reset the lap timer.
DISABLE VTX CONTROL	Disables controlor
LAUNCH CONTROL	Selve: merlces oo0185. PLLB Hagor Gent0 Plin hyng
MSP OVERRIDE	STICK COMMANDS DISABLE

Settings and description of modes:

Ranges

Activate when the receiver channel matches the specified input values. A receiver channel that gives a reading between a range min/max will activate the mode.

Links

Activate when another linked Mode is active.

Multiple Ranges, Links can be matched, combined using boolean AND or OR operators to combine activation conditions for a Mode.

13.1.15 BETAFlight - Motors Tab

Configure motor and ESC settings. Change motor direction and mixing, and set up advanced telemetry and flight features.

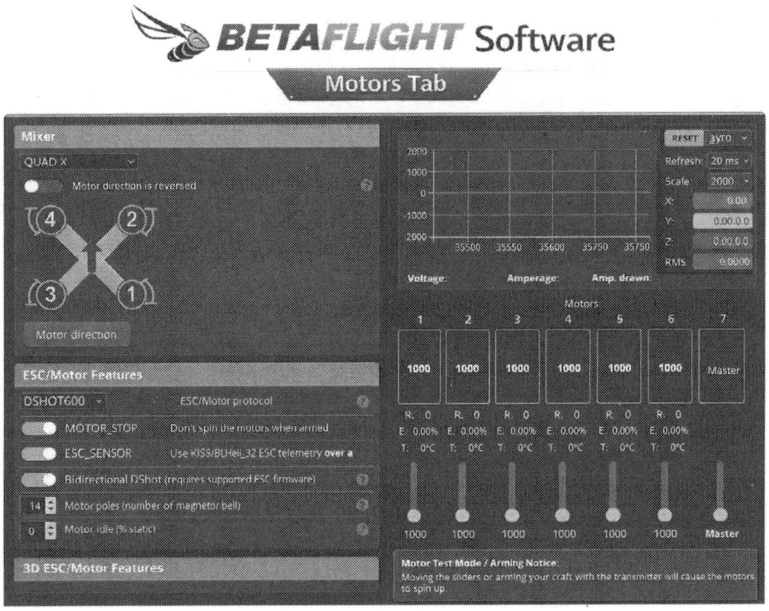

Mixer

The Mixer defines the motor layout and determines how the flight controller (FC) drives each motor to maintain stable flight. For most quad-copters, the standard and recommended option is QUAD X. If you are unsure which layout to use, select QUAD X, as it applies to the vast majority of modern quad builds.

Motor Direction Is Reversed

This setting controls the direction in which the motors spin relative to the frame:

Normal (Props In)

- The default configuration. Front motors spin inwards toward the camera.

Reversed (Props Out)

- Motors spin outwards away from the camera.

Why pilots use "props out":

- Reduces the chance of debris (dust, grass, sand) being thrown into the FPV camera

- Can slightly reduce prop wash on smaller or lighter aircraft

Trade-off:

Debris is instead pushed toward the body of the quad, which may increase wear on electronics and frame components.

Motor Direction Tool

This opens the Motor Direction Test Tool, allowing you to:

- Gently spin each motor individually
- Verify correct motor order
- Reverse motor direction electronically (without re-soldering wires)

Safety warning:

Always remove propellers before using this tool. Spinning motors with props installed is extremely dangerous.

ESC / Motor Features - ESC / Motor Protocol

For modern builds, DShot is the standard and recommended protocol. It offers:

- Digital signal transmission (no calibration required)
- Faster and more accurate motor response
- Advanced features such as RPM filtering and telemetry

Recommended DShot pairings:

- 8 kHz PID loop → DShot600
- 4 kHz PID loop → DShot300
- 2 kHz PID loop → DShot150

Slower DShot speeds cannot send updates fast enough to fully benefit from higher PID loop frequencies, which is why correct pairing is important.

Legacy protocols:

Older protocols such as Oneshot125 are only required for very old ESCs (e.g. original BLHeli). Modern ESC firmware such as BLHeli_S, BLHeli_32, BlueJay, or AM32 should always use DShot.

MOTOR_STOP

When enabled, MOTOR_STOP prevents motors from spinning at idle when the quad is armed.

- Disabled (recommended): Motors spin slowly at idle, improving stability and responsiveness.
- Enabled: Motors remain stopped until throttle is applied. This is generally not recommended and is rarely needed.

In summary, correct Mixer, motor direction, and ESC protocol configuration is essential for safe and predictable flight. Always verify motor order and spin direction before installing propellers or attempting a maiden flight.

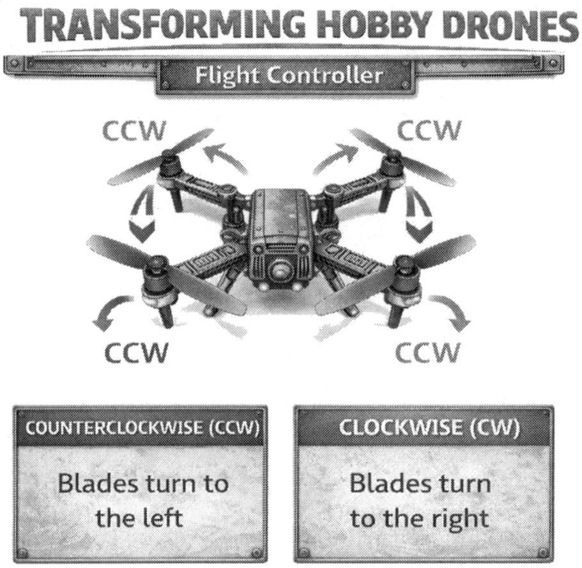

13.1.16 BETAFlight - OSD Tab

Betaflight OSD allows the pilot to view in-flight information and telemetry data overlaid onto their FPV camera feed. This is where you can configure the OSD elements and their layout, as well as the alarms and warnings which will be displayed.

Beat Flight Software - OSD (On Screen Display) Tab

The OSD (On-Screen Display) feature requires a supported device to overlay real-time flight data onto the video feed. Supported devices include the onboard MAX7456 chip found on many analogue flight controllers, the FrSky OSD board, and compatible digital video transmission systems that support native OSD integration. These systems allow telemetry data—such as battery voltage, current draw, RSSI/LQ, GPS information, altitude, flight mode, and warnings—to be rendered directly into the pilot's video stream.

In analogue systems, the MAX7456 chip inserts OSD data into the composite video signal before transmission, providing a lightweight and reliable solution widely used in hobby and operational drones. In digital FPV systems, OSD data is typically injected via a digital protocol and rendered within the headset or display, offering higher resolution, improved clarity, and greater flexibility in layout and scaling.

Proper OSD configuration is critical for both safety and mission effectiveness, as it provides the pilot with immediate situational awareness and system health information without diverting attention from the aircraft's visual feed.

OSD Elements & Layouts , MSP Displayport protocol.

OSD elements are enabled by toggling the checkbox next to each item. Once enabled, elements can be re-positioned by dragging them within the Preview pane. Any changes are reflected in real time and can be viewed immediately through the VTX and goggles.

- Three OSD profiles are available
- Profiles can be switched in flight or on the ground
- Each profile can have different elements and layouts (useful for racing, cruising, or diagnostics)

Active OSD Video formats

The Active OSD Video Format setting determines how on-screen display information is rendered and which layout profile is overlaid onto the live camera feed. This selection controls the OSD grid structure and must match the video system being used to ensure correct alignment and readability. The Auto option attempts to automatically detect the appropriate video format, reducing the need for manual configuration. PAL uses a 30-column by 16-row grid and, when a MAX7456 OSD chip is in use, must match the camera's video signal to prevent distortion or misaligned elements. NTSC operates on a 30-column by 13-row grid and carries the same requirement when using the MAX7456. The HD format is designed specifically for digital VTX systems and communicates OSD data via the MSP Displayport protocol, allowing higher-resolution layouts that are dynamically managed by the digital video system.

- Boot Logo
- A custom boot logo can be uploaded. The image must be 288x72 pixels, and green/white/black. Green areas are rendered as transparent. Use Upload Font to save changes
- Upload Font
- Font upload max be required when upgrading from older versions of Betaflight as font symbols have been added and changed over time. In particular if your cross-hair looks wrong or if your display appears garbled or using incorrect characters you should try uploading the font again.

HD OSD behaviour:

By default, HD OSD uses a 53 × 20 grid. When the VTX comes online, Betaflight queries it via MSP Displayport to determine the optimal grid size and may automatically adjust the layout to match what the digital system supports.

Unit

The Units setting defines the measurement system used to display telemetry and flight data on the OSD, allowing pilots to tailor information to their regional standards or personal preference. When Metric is selected, speed is shown in kilometres per hour, distance is measured in kilometres, and temperature is displayed in degrees Celsius. The Imperial system displays speed in miles per hour, distance in miles, and temperature in degrees Fahrenheit, which is commonly used in the United States. The British setting provides a hybrid format, showing speed in miles per hour while retaining kilometres for distance and Celsius for temperature. Selecting the appropriate unit system ensures that flight data is immediately understandable at a glance, reducing cognitive load and improving situational

Timers

In BETAFlight, the Timers feature allows pilots to configure up to three independent timer alarms for on-screen display monitoring. Each timer can be assigned a specific source—such as total flight time, time since arming, or throttle-based time—and an alarm threshold value. When the selected timer reaches its defined limit, an OSD alert is displayed to notify the pilot. This provides a clear and reliable way to manage battery usage, mission duration, and overall flight safety without relying on external

Alarms

In BETAFlight, the Alarms feature is used to display on-screen display alerts when monitored flight parameters reach or exceed predefined threshold values. These parameters commonly include critical metrics such as battery voltage, current draw, or other sensor-based readings. When a threshold is crossed, a clear OSD alarm is triggered, immediately drawing the pilot's attention to a potential issue. This early warning system is essential for preventing over-discharge of batteries, avoiding power-related failures, and maintaining safe control of the aircraft during flight.

Warnings

In BETAFlight, the Warnings feature provides real-time on-screen display alerts that indicate error states or important system conditions detected by the flight controller. These warnings can originate from multiple sources, such as low battery voltage, failsafe activation, GPS status problems, or other critical system events. The warnings are designed to immediately inform the pilot of issues that may affect flight safety or performance. For pilots who prefer a less cluttered display, individual warning types can be selectively disabled, allowing a balance between situational awareness and a clean, unobstructed OSD layout.

Post Flight Statistics

In BETAFlight, the Post-Flight Statistics feature provides a summary screen on the OSD immediately after the craft disarms. This screen displays selected flight metrics such as total flight time, maximum speed reached, battery voltage sag, distance travelled, and other relevant performance data. By presenting this information at the end of each flight, pilots can quickly assess battery health, flight efficiency, and overall system performance without needing to review logs or external tools, making it a valuable aid for both tuning and operational awareness.

OSD Font Manager

The BETAFlight OSD Font Manager is used to manage and upload the character set (font glyphs) displayed on the on-screen display when using an analog OSD chip such as the MAX7456. These glyphs define how text, numbers, icons, and symbols—such as voltage indicators, crosshairs, and warning icons—appear in the video feed. The Font Manager allows pilots to upload updated or custom fonts directly to the flight controller, ensuring compatibility with the current BETAFlight version. Re-uploading the font is especially important after firmware upgrades, as new symbols may be added or existing ones changed; symptoms like garbled text, incorrect characters, or a distorted crosshair are common indicators that a font upload is required. Proper font management ensures the OSD remains clear, readable, and functionally accurate during flight.

BOOT Logo

The BETAFlight OSD Boot Logo feature allows pilots to upload a custom image that is displayed on the on-screen display during system startup. This logo appears before normal OSD elements are shown and can be used for branding, identification, or simple visual customisation. The boot logo must meet specific requirements: a resolution of 288 × 72 pixels and a colour palette limited to green, white, and black. Within the image, green areas are treated as transparent, allowing the underlying video to show through. Once the logo is prepared, it is uploaded using the Upload Font function, which saves both the boot logo and any associated font changes to the flight controller, ensuring they are applied correctly on startup.

Upload Font

In BETAFlight, OSD font uploads may be required after upgrading the firmware, as symbols and glyphs can change between versions. When the stored font does not match the current firmware, visual issues can occur, such as an incorrect or distorted cross-hair, garbled characters, or missing icons and symbols on the display. In these cases, re-uploading the OSD font ensures that the character set is fully compatible with the installed BETAFlight version, restoring correct rendering and maintaining a clear, reliable on-screen display during flight.

In summary, the OSD tab allows full control over what flight data is shown, how it is displayed, and how it behaves across analog and digital video systems, making it a critical tool for both usability and situational awareness during flight.

13.1.17 BETAFlight - PID Tuning Tab

In BETAFlight, the PID Tuning tab is used to configure the PID controller, filter settings, and rate values, which together define how the aircraft responds to pilot inputs and stabilises itself in flight. Betaflight uses a profile system that allows multiple configurations to be saved and quickly switched, making it easy to tailor the flight behaviour for different conditions, aircraft setups, or pilot preferences. On the top left of the PID Tuning tab, the active profile can be selected for editing.

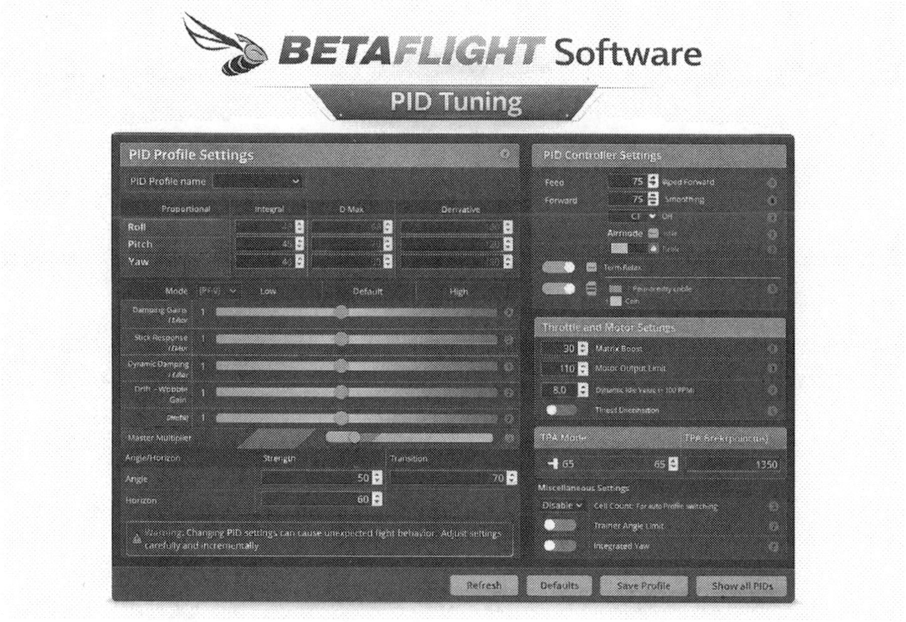

There are four PID profiles, each capable of storing independent PID and filter settings, and four rate profiles, which store independent rate and expo configurations. This separation allows pilots to adjust flight responsiveness without altering core stabilisation behaviour, or vice versa. On the top right of the configurator, additional management options are available, including Copy Profile, which duplicates the current PID and filter settings to another profile; Copy Rate Profile, which copies the current rate configuration to another rate profile; and Reset This Profile, which restores the selected profile to default values. The Show All PIDs option reveals additional PID values, including those related to the magnetometer, providing advanced users with deeper tuning visibility when required.

- Copy profile: Make a duplicate of the current profile to another profile.
- Copy rateprofile: Make a duplicate of the current rate profile to another rate profile.
- Reset this Profile: Reset the current profile to default values.
- Show all PIDs: Show PID values for the Magnetometer.

Betaflight allows to tune PIDs using simplified tuning. The sliders can be used to adjust the PID value indirectly. When the Mode is set to OFF, the sliders are disabled, and the user can input the PID values directly.

Mode

- OFF: no sliders, enter values manually
- RP: sliders control Roll and Pitch only, enter Yaw values manually
- RPY: sliders control all PID values

[WARNING]

Going from RP to RPY mode will overwrite Yaw settings with firmware settings.

Sliders

Sliders to adjust the aircraft flight characteristics (PID gains).

In BETAFlight, the PID Tuning Tab sliders provide an intuitive way to shape flight behaviour by adjusting multiple underlying parameters simultaneously, without needing to manually tune individual PID values. The Master Multiplier acts as a global scale for the entire PID system, increasing or reducing overall control authority and making the quad feel either more aggressive and locked-in or softer and more forgiving. Damping (D gains) primarily influences the Derivative term, helping to reduce oscillations, prop wash, and bounce-back by smoothing rapid movements, though excessive damping can increase motor heat and noise. Tracking (P & I) adjusts how closely the craft follows both immediate and sustained commands, with Proportional gain improving responsiveness to quick errors and Integral gain enhancing long-term stability, especially during steady flight or in windy conditions.

The Stick Response slider controls how directly the aircraft reacts to pilot inputs, affecting how sharp or smooth the quad feels around centre stick and during fast manoeuvres. Dynamic Damping adapts D-term behaviour based on flight conditions, increasing damping only when needed to maintain smoothness while minimising unnecessary motor stress. The Drift setting influences how strongly the quad resists unwanted movement over time, helping it maintain heading and attitude without slowly wandering. Pitch and Pitch Damping allow axis-specific tuning, adjusting how aggressively the aircraft responds on the pitch axis and how smoothly that movement is controlled, which is particularly important for forward flight, dives, and throttle transitions. Together, these sliders offer a powerful yet accessible way to fine-tune flight performance, balancing responsiveness, stability, and efficiency across different flying styles and conditions.

DIP- Controller Settings

Feed forward

In BETAFlight, the PID Tuning tab sliders allow pilots to fine-tune how the aircraft responds to stick inputs and how smoothly that response is translated into motor output, without directly editing individual PID values. Feed-forward controls how strongly the quad reacts to pilot stick movement itself rather than waiting for an attitude error to develop, resulting in a more immediate and connected feel. Boost enhances feed-forward during fast stick movements, sharpening snap manoeuvres and improving responsiveness, while Transition manages how feed-forward behaviour blends from low to high stick input, keeping control smooth near centre while remaining aggressive at the extremes. The Max Rate Limit caps the maximum rotational speed commanded by feed-forward, preventing excessive or uncontrollable rotations.

Jitter Reduction,

Smoothness, and Averaging work together to manage noise and micro-oscillations in both stick input and gyro data. These settings help reduce twitchiness and high-frequency vibration at the cost of a small increase in latency, making them especially useful on noisier builds or long-range setups. Pitch Damping (Pitch & Roll) provides axis-specific damping to control overshoot and bounce-back during aggressive pitch or roll manoeuvres, improving stability in dives, flips, and rapid direction changes. Collectively, these sliders allow pilots to balance precision, smoothness, and control authority, tailoring the quad's flight characteristics to racing, freestyle, cinematic, or long-range flying styles.

Other, critical settings

In BETAFlight, the PID Tuning tab sliders provide high-level control over how the flight controller manages stability, throttle changes, and rapid manoeuvres. I-Term Relax limits how aggressively the Integral term responds during fast stick movements, preventing I-term wind-up that can cause bounce-back after flips and rolls. Anti-Gravity temporarily boosts I-term during rapid throttle changes, helping the aircraft maintain attitude and resist sudden drops or climbs when throttle is punched or cut. I-Term Rotation improves consistency during combined axis movements by allowing the I-term to rotate with the craft, maintaining stable control during complex manoeuvres.

The Dynamic Damping Gain slider controls how strongly D-term damping is applied in response to changing flight conditions, increasing stability only when needed to reduce oscillations and prop wash while minimising unnecessary motor stress. Dynamic Advance enhances responsiveness during rapid stick inputs by anticipating required corrections, improving precision and reducing delay without sacrificing smoothness. Damping (Pitch & Roll) allows axis-specific adjustment of damping strength, enabling finer control over pitch and roll behaviour during aggressive manoeuvres. The Transition setting governs how these dynamic effects blend between low and high stick inputs, ensuring smooth, predictable handling near centre stick while retaining strong control authority at higher deflections.

Throttle and Motor Settings

In BETAFlight, the PID Tuning tab sliders also include several settings that directly influence how throttle input is translated into motor output and how the aircraft behaves under changing power conditions. Throttle Boost momentarily increases motor output in response to rapid throttle changes, improving punch-outs and helping the quad feel more responsive during quick climbs or throttle recoveries. The Motor Output Limit caps the maximum motor command sent to the ESCs, which can be used to reduce overall power for efficiency, manage overheating, or comply with power restrictions while preserving resolution across the throttle range.

The Dynamic Idle Value sets the minimum motor speed when armed, adjusting idle RPM dynamically to prevent motor stalling during rapid descents, sharp turns, or aggressive braking manoeuvres. VBAT Sag Compensation counteracts the natural voltage drop that occurs under load as the battery discharges, maintaining a more consistent throttle response throughout the flight. Thrust Linearisation reshapes the throttle curve so that changes in throttle input result in more predictable and linear changes in thrust, making fine throttle control easier and improving consistency across different battery voltages and load conditions.

Miscellaneous Settings

In BETAFlight, the PID Tuning tab sliders also include several settings that affect overall control logic, safety limits, and how the flight controller interprets pilot intent. Cell Count allows the flight controller to correctly identify the number of battery cells in use, ensuring that voltage-based features such as VBAT sag compensation, alarms, and thrust linearisation operate accurately. Acro Trainer Angle sets the maximum tilt angle permitted when Acro Trainer mode is enabled, providing a safety boundary that helps prevent over-rotation while still allowing manual control, making it useful for skill development and recovery from disorientation.

Integrated Yaw blends yaw control into coordinated roll and pitch movements, improving heading stability and creating smoother, more natural turns, particularly during forward flight. Absolute Control helps maintain a consistent aircraft attitude relative to pilot input, reducing drift and unintended rotation when the sticks are centred. This feature improves precision and predictability, especially during technical manoeuvres or when returning to level flight, by ensuring the quad follows the pilot's intended orientation rather than gradually deviating over time

DIP - Tunning

The PID Tuning settings in BETAFlight control how the flight controller stabilises the aircraft and responds to pilot inputs by continuously correcting errors in attitude and movement. These settings are based on the PID control loop, which consists of Proportional (P), Integral (I), and Derivative (D) terms. The P gain determines how strongly the craft reacts to immediate errors, affecting overall responsiveness and firmness. The I gain corrects long-term or persistent errors, helping the quad hold its angle during sustained manoeuvres or in windy conditions. The D gain reacts to rapid changes in error, damping oscillations and smoothing movement, but excessive D can introduce noise and heat in the motors.

Alongside the core PID values, the PID Tuning tab also includes filter settings, which are used to remove unwanted vibration and electrical noise from gyro and motor signals without introducing excessive latency. Proper filtering is essential for clean sensor data and stable flight, especially on high-performance builds. Rate settings, configured separately via rate profiles, define how fast the aircraft rotates in response to stick inputs and how sensitive the controls feel around centre stick. Together, the PID values, filters, and rates form the foundation of a quadcopter's flight characteristics, and careful tuning allows pilots to balance stability, smoothness, and agility to suit different flying styles and environments

13.1.18 BETAFlight - Video Transmitter Tab

In BETAFlight, the Video Transmitter (VTX) tab is used to configure the settings of a supported video transmitter directly from the flight controller. This tab allows the pilot to set key parameters such as output power level, frequency band, channel, and other VTX-related options without needing to use physical buttons or separate configuration tools. Centralising VTX control within Betaflight simplifies setup, ensures settings are correctly applied before flight, and makes it easy to change channels or power levels to avoid interference or comply with local regulations.

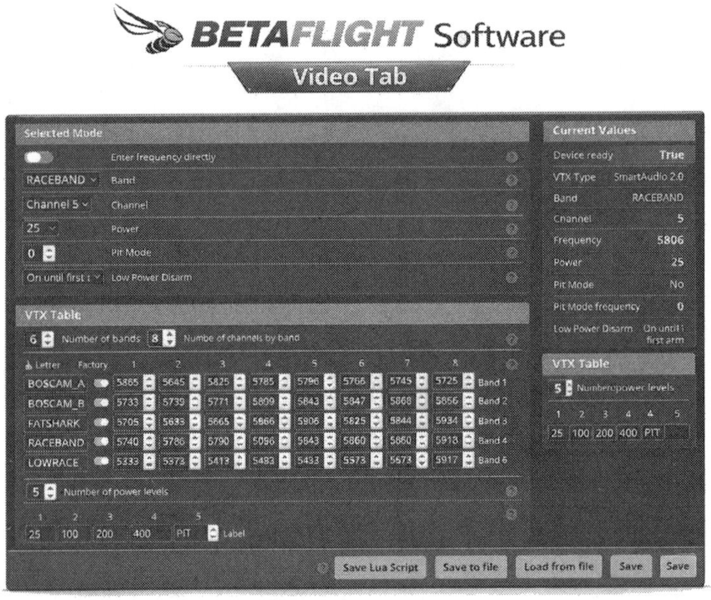

Note: When using DJI or Walksnail digital video transmitters, this tab can generally be ignored, as those systems are configured through their own software and do not rely on Betaflight's VTX control. The Betaflight VTX tab is primarily intended for analogue VTXs and HDZero systems, where direct configuration via the flight controller is required and supported.

Selected Mode

In BETAFlight, the functionality described here belongs to the Video Transmitter (VTX) tab, not the PID Tuning tab. The Selected Mode section of the VTX tab determines how video transmitter settings are applied and controlled by the flight controller. Pilots can choose to enter a frequency directly, which is useful for precise tuning or for operating outside standard band and channel presets where regulations permit. Alternatively, selecting a predefined band and channel simplifies configuration and ensures compatibility with common FPV frequency plans. The Power setting controls the VTX output level, allowing quick adjustment between low power for bench testing or shared flying environments, and higher power levels for increased range and signal penetration.

Additional safety and interference-reduction features are also managed in this tab. Pit Mode drastically reduces transmission power while the quad is powered but not flying, preventing disruption to other pilots during setup. When enabled, a Pit Mode Frequency can be specified so the VTX operates on a safe, low-impact frequency while grounded. The Low Power Disarm option automatically forces the VTX into a low-power state whenever the aircraft is disarmed, reducing heat buildup and unnecessary RF emissions. Together, these settings provide precise control over video transmission behaviour, improving safety, regulatory compliance, and usability in both solo and multi-pilot environments.

The VTX Table defines the capabilities of the connected video transmitter so the flight controller knows exactly which bands, channels, frequencies, and power levels it can control. It specifies the number of supported bands and the number of channels per band, with each band identified by a letter name (such as A, B, E, F, or R) and mapped to exact channel frequencies. The table also defines the number of available power levels and their corresponding output values, typically in mills-, ensuring that selections made in the VTX tab accurately match the hardware's real capabilities.

At the bottom of the VTX Table are management tools that simplify configuration and reuse. Save writes the current VTX table directly to the flight controller, Save to File exports it for backup or sharing, and Load from File restores a previously saved configuration. Load from Clipboard allows rapid pasting of VTX table data from documentation or community presets. Finally, Save Lua Script generates an "mcuid.lua" file for use with Betaflight transmitter Lua scripts; version 1.6.0 and above can use the file directly, while older versions require the file to be renamed to match the model name on the transmitter.

13.1.19 BETAFlight - GPS Tab

In BETAFlight, the GPS tab is used to configure and monitor all GPS- and GNSS-related functions, enabling features such as position hold, rescue modes, telemetry, and enhanced situational awareness. The GPS Configuration section defines how the flight controller communicates with the GPS module. The Protocol setting selects the communication standard used by the GPS (such as UBLOX, NMEA, or MSP), ensuring compatibility between the module and the flight controller. Auto Configuration, when enabled, allows Betaflight to automatically configure supported GPS units for optimal operation, reducing manual setup errors. The Use Galileo option enables additional satellite constellations alongside GPS, improving accuracy, fix speed, and reliability, particularly in challenging environments. Set Home Point Once locks the home position after the first valid fix, preventing it from being reset mid-flight, which is important for accurate return-to-home and distance calculations. Ground Assistance Type configures how auxiliary data is used to stabilise GPS performance when the craft is near the ground, while Magnetometer Declination compensates for regional magnetic variation, ensuring accurate heading information when a magnetometer is used.

The GPS Options and Status area provides real-time feedback on GPS performance and lock quality. Indicators such as 3D Fix confirm that the GPS has achieved a full positional solution, while the number of satellites shows how many are currently contributing to that fix. Altitude, speed, and heading data are displayed using GPS and IMU fusion, allowing accurate tracking of movement and orientation. The current latitude and longitude are shown continuously, along with distance to home, which is critical for navigation and rescue features. Positional PDOP (Position Dilution of Precision) is also displayed, providing a numerical indicator of fix quality—lower values represent higher positional accuracy.

For deeper diagnostics, the GPS tab also presents detailed GNSS signal information. This includes GPS signal strength data for individual satellites, the GNSS ID identifying which satellite system is in use, Satellite ID (Sat ID) numbers, and per-satellite signal strength values. Each satellite's status and quality indicate whether it is being tracked or actively used in the position fix. Together with the continuously updated current GPS location, these diagnostics allow pilots to assess GPS health, troubleshoot reception issues, and confirm that the navigation system is performing.

13.1.20 BETAFlight - Other Tips & Comments

BETAFlight – Ongoing Development Goals (with focus on AI & Machine Learning)

BETAFlight is an open-source flight-control firmware project focused on delivering fast, reliable, and highly configurable control for multi-rotor aircraft. Its ongoing development goals prioritise flight performance, robustness, accessibility, and safety, with emerging interest in automation, adaptive control, and data-driven tuning—areas that increasingly intersect with concepts from AI and machine learning (ML), even if not always labelled as such.

Core Development Direction:

- The primary goals of Betaflight remain:
- Lower latency and higher control fidelity
- Improved noise handling and filtering
- Better defaults that work “out of the box”
- Reduced tuning complexity for pilots
- Stable operation across a wide range of hardware

These goals drive continual refinement of the control loop, filtering pipeline, motor protocols, and sensor fusion.

AI & Machine Learning-Adjacent Development

While Betaflight does not currently embed heavyweight AI or neural networks onboard (due to MCU processing, memory, and real-time determinism constraints), many ongoing developments align closely with machine-learning principles:

Adaptive and Dynamic Control Systems

Features such as dynamic filtering, dynamic idle, dynamic damping, anti-gravity, and feedforward modelling adjust behaviour in real time based on flight conditions. These systems mimic ML outcomes by responding to patterns in data, even though they are implemented using deterministic algorithms rather than trained models.

Data-Driven Tuning Philosophy

Blackbox logging, RPM filtering, and spectral analysis tools enable pilots (and developers) to analyse large datasets of real flight behaviour. This supports offline optimisation, where ML tools external to Betaflight can already be used to identify optimal PID and filter configurations—feeding results back into firmware settings.

Automatic Configuration & Smarter Defaults

Betaflight increasingly focuses on auto-configuration (e.g. gyro alignment, ESC protocol detection, GPS setup). The long-term aim is to reduce manual tuning by encoding learned best-practice into firmware defaults—conceptually similar to trained model behaviour, but implemented safely and predictably.

AI Constraints in Real-Time Flight Control

True onboard AI/ML faces major constraints in Betaflight's operating environment:

- Hard real-time deadlines (PID loops at 2–8 kHz)
- Limited MCU resources
- Safety-critical determinism requirements
- Certification and predictability concerns

As a result, developers prioritise explainable, deterministic control logic over opaque models—especially for low-level stabilisation.

Likely Future Trajectory

Looking forward, Betaflight's AI-related evolution is most likely to occur through:

- External AI-assisted tuning tools (PC-side analysis of Blackbox logs)
- Adaptive algorithms inspired by ML, but implemented deterministically
- Improved sensor fusion using smarter weighting of IMU, GPS, barometer, and magnetometer data

- Context-aware flight behaviour, where firmware dynamically optimises response based on environment, battery state, and flight style
- Integration with higher-level autonomy stacks (e.g. companion computers running AI vision or navigation), while Betaflight remains the real-time stabilisation layer

Summary

In essence, Betaflight's ongoing development does not aim to replace its control system with neural networks, but rather to absorb the benefits of AI thinking—adaptivity, optimisation, and data-driven refinement—while preserving real-time safety and transparency. AI and ML are therefore more likely to enhance how Betaflight is tuned, configured, and supported, rather than directly replacing its core flight-control logic.

Appendix (A) - Hobby Drone Regulations

For the specific regulations for your region, please reference Appendix Below, (Drone Regulations listed by Country)

[IMPORTANT}

Drone regulations vary significantly by country, with many countries like Australia, the UK, and the US having detailed rules, while others have less formal restrictions. Key common regulations include flight height limits (often 120m), keeping drones within visual line of sight, restrictions on flying over crowds or sensitive areas (like airports and military sites), and different rules for commercial use which often require registration or a license.

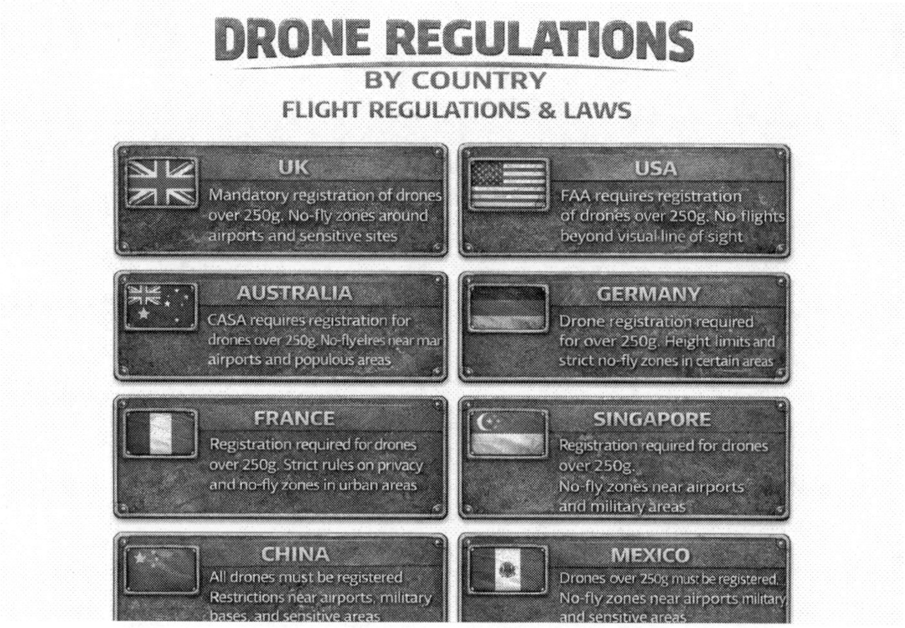

Australia

Governed by Civil Aviation Safety Authority (CASA). Recreational: Fly below 120m, stay 30m away from people, and use a CASA-verified drone safety app to check restrictions. Commercial: Requires a Remote Pilot Licence (RePL) or an Operator's Certificate (OC). Drones weighing over 250g commercial use require registration.

China:

Registration with local authorities and compliance with flight bans near government agencies.

France:

Bans on flying near airports and restrictions on violating privacy.

Germany:

Registration and a license are required for drones over 250g.

Maldives:

Permission from local authorities is required, along with respect for bans on flying over crowds and residential areas.

Mexico:

Commercial use is restricted to registered Mexican businesses, and authorities have been known to stop tourists even for recreational flights.

Singapore:

An operator permit is required for non-recreational use, and other permits may be needed from the IMDA or SPF.

Turkey:

Registration and permits for drones over 500g, with flight bans near military installations.

United Kingdom:

Recreational: Fly below 120m and 500ft AGL, and at least 50m away from people or property.
Commercial: CAA approval is required.

United States: Governed by the FAA:.

Recreational: Fly below 400ft AGL in Class G airspace.
Commercial: Requires a Part 107 Remote Pilot license.

UAE:

Special permission for night flights, and respect for no-fly zones near airports.

General rules to follow

- Fly during the day: Many countries require drones to be flown during daylight hours only.
- Keep it in sight: Always maintain visual line of sight with your drone at all times.
- Respect restricted areas: Do not fly over or near airports, government facilities, power plants, or emergency operations.
- Use a safety app: Many countries recommend or require using a drone safety app to find where you can and cannot fly.
- Understand your drone's weight: Be aware that rules often differ for drones under and over a certain weight, typically 250g.
- Be aware of privacy: Respect the privacy of others and avoid flying over private property or crowded

areas.

- Commercial vs. Recreational: The rules for commercial drone use are generally much stricter and often require specific licenses or registrations

Appendix (B) - Deployment of Hobby Drones

From the initial stages of the full Russian invasion of Ukraine in 2022, Peter Savage volunteered on the ground to support aid missions flowing from Australia, the United States, and the United Kingdom into Kyiv. Over the course of 36 months of involvement in the country, including repeated deployments and visits to front-line areas to deliver humanitarian assistance, he was repeatedly encouraged by the Ukrainian people he met to seek greater support from Ukraine's allies—particularly in the provision of drones. This support included both semi-commercial systems and hobby-grade drone kits, which were increasingly critical to front-line defence and civilian protection. Drawing directly on what he witnessed, learned, and experienced firsthand, this book has been written as a testament to those experiences, documenting the realities on the ground and the vital role that accessible drone technology came to play in defending the country and its people.

Example 1 - Highly Mobile Drone Recon Unit (Quad-Copter)

The examples illustrates a university-based drone unit that emerged shortly after the full-scale invasion of Ukraine, catalysed by the shock of witnessing the Bucha massacre. Initially formed by just three or four students, the group drew on their existing knowledge of hobby drones, consumer technology, and personal equipment to establish a small militia-style capability. The unit was soon attached to an infantry formation, beginning with reconnaissance and observation tasks, but over the course of approximately eighteen months it evolved into a combat-capable drone unit.

In its early phase, the team relied on basic DJI drones fitted with long-range, high-gain antennas and simple radio-controlled interfaces, operating at a time before the widespread deployment of dedicated counter-drone systems. As electronic warfare and anti-drone technologies became more prevalent, the unit was forced to adapt rapidly, modifying hobby-grade platforms to improve resilience against FM interference and electronic attack. Survival depended not only on technical innovation, but also on mobility and disciplined tactics: operating from the rear of a 4×4 vehicle, deploying quickly, completing missions within minutes, and relocating before enemy artillery or direction-finding systems could detect and engage their position. This blend of ingenuity, adaptability, and tactical awareness allowed the unit to remain operational despite escalating threats.

By late 2023, the team had grown significantly in size and capability, establishing a small improvised laboratory hidden within an abandoned block of flats, building drones similar to the DIY project described in this book. Using basic carbon-fibre frames and a modular design philosophy, they developed drones tailored to specific missions and environmental conditions, allowing sensors, payloads, and components to be swapped or replaced rapidly in the field.

By the end of the year, the unit had become so effective that many of the original student members progressed into leadership roles, forming and running their own drone teams while coordinating with international suppliers and developers. Building on this success, the group began developing fixed-wing drones to complement their quad-copter fleet. While their early reliance on hobby kits proved effective, it also exposed a lack of aviation and flight-dynamics knowledge—an issue often overlooked by quad-copter developers transitioning to fixed-wing platforms. Many early test flights ended in crashes shortly after take-off due to poor payload balance, misjudged launch technique, or limited understanding of wind and flight conditions.

Recognising this limitation, the team rapidly adapted by incorporating basic aviation skills into their training, establishing a simple but effective flight syllabus to address these shortcomings. Even fundamental lessons —such as launching into the wind to maximise lift—produced immediate improvements. As their confidence and competence grew, a broader range of drones began to emerge, developed entirely within Ukraine and increasingly focused on the integration of artificial intelligence. Alongside this practical development, the team built a growing academic understanding of enemy tactics, using AI-assisted analysis to identify opposing units by deployment patterns, infrared signatures, acoustic profiles, and the physical silhouette or cross-section of vehicles in the field.

The author came to understand that what began as a small group of young men using hobby-grade drones to defend their country in 2022 had, within a few years, evolved into one of the leading forces in drone innovation. By 2026, this group had grown into a large, highly capable team contributing not only to front-line drone operations but also to the development of anti-drone technologies in Ukraine and beyond. Their journey—from an improvised collective to a technologically sophisticated organisation integrating artificial intelligence and machine learning into modern warfare—stands as a powerful example of how necessity, education, and adaptation can reshape the future of conflict.

Example 2 - Battalion Combat Unit, operating from the (Trenches of Kherson)

This photograph was taken just before the liberation of Kherson in early November 2022, Peter Savage had already travelled to the front-lines of the war in Ukraine to deliver humanitarian aid, only to find himself drawn into the trenches around Kherson, where Ukrainian forces demonstrated their most urgent needs: not only food, warmth, and water, but also drone technology, counter-drone equipment, and 3D printers. The image captured by Peter shows a battalion drone commander loading a modified grenade beneath a DJI drone, preparing it to fly roughly a thousand metres east toward Russian positions before releasing its payload.

This and several subsequent sorties proceeded smoothly, each lasting just over twenty minutes. Although the grenades were quick to load, they appeared fragile, having been adapted with 3D-printed tail fins and modified impact fuses to ensure detonation on contact. At the time, the process was simple and streamlined, something Nikolai, the pilot, carried out with confidence and ease. Each sortie achieved an estimated ninety percent success rate, with the result being the neutralisation of two or three young Russian recruits per hour —highly effective given the drone's operational envelope of up to two kilometres in altitude and roughly eight kilometres inland. It was important to understand that, for the Ukrainians, the concept of "aid" was intrinsically broad, encompassing civilian necessities alongside defensive and tactical technologies. Visiting the front lines to deliver assistance inevitably became a guided exposure to the areas of greatest suffering, ensuring that the most urgent needs were made visible to allies capable of helping to meet them.

Example 3 - Electronic Warfare Unit, Kyiv

This example of an eight-rotor drone was developed in a discreet laboratory in Kyiv toward the end of 2023 to meet a highly specific mission profile requiring an exceptional lift-to-weight ratio and the ability to operate quietly over significant distances behind enemy lines.

Large custom Multi-Rotor Drone - Diameter three metres

Once deployed, the platform was used to covertly insert a camouflaged piece of equipment that remained concealed for months, operating remotely in a sophisticated electronic warfare role. The system interfered with enemy command-and-control networks, assisted in target identification, and enabled the hijacking of hostile systems, working in conjunction with continued drone sorties that communicated through an ongoing mesh protocol with other active missions. Employed on multiple occasions, the drone—constructed at a cost of approximately USD 3,000—inflicted hundreds of thousands of dollars' worth of damage before being deliberately self-destructed upon discovery, thereby protecting the secrecy, development, and technical sophistication of the capability involved.

Example 4 - Fixed-Wing Drone, long duration missions in the Donbas, Oblast.

This example of a fixed-wing reconnaissance drone deployed in the Donbas region of Ukraine illustrates how a modified hobby-grade air-frame can be transformed into a highly capable medium-range intelligence, surveillance, and reconnaissance (ISR) platform. Built around a lightweight balsa-wood core coated with a thin carbon-fibre composite skin, the air-frame balances structural rigidity with minimal mass, allowing for a wingspan of approximately two metres while maintaining a low wing loading. This configuration provides excellent lift-to-drag characteristics, enabling efficient cruise flight, stable loitering behaviour, and predictable stall margins—critical factors for extended missions conducted far beyond friendly lines. The aerodynamic profile was optimised for endurance rather than speed, favouring high aspect-ratio wings, gentle camber, and slow-flight stability to maximise time on station above enemy positions.

Powering the platform is a hybrid energy system that combines next-generation solid-state lithium batteries (Li-SSB) with lightweight solar panels integrated into the upper wing surfaces. The use of Li-SSB technology significantly improves energy density and thermal stability when compared to conventional lithium-polymer cells, reducing fire risk while extending total flight time. Solar augmentation, while modest in raw output, provides a continuous trickle charge during daylight operations, effectively flattening the discharge curve during cruise and loiter phases. This combination allows the aircraft to maintain sustained flight at operational altitudes exceeding 2,000 metres above ground level, dramatically increasing its loiter time and reducing detectability from ground-based observers and small-arms fire. At these altitudes, the aircraft can exploit thinner air for improved aerodynamic efficiency while remaining within the optimal performance envelope of its electric propulsion system.

To mitigate the risk of loss when targeted by Russian electronic warfare assets, the drone was fitted with one of the early operational artificial intelligence modules capable of autonomous decision-making under attack. When exposed to radio-frequency jamming, GPS spoofing, or command-link disruption, the system could transition into an independent stealth mode. In this state, the drone disengaged from direct operator input and executed pre-learned or dynamically generated flight profiles designed to minimise detectability and preserve mission integrity. Rather than relying solely on GNSS data, the onboard AI fused inputs from inertial measurement units (IMUs), barometric altimeters, magnetometers, optical flow sensors, and terrain-referenced visual navigation. This sensor fusion allowed the aircraft to maintain situational awareness, stabilise its flight path, and continue reconnaissance tasks even in contested electromagnetic environments.

Equipped with a compact but sophisticated sensor suite, the drone proved highly effective in identifying personnel movements and vehicle activity on the ground. Electro-optical cameras, combined with infrared imaging, enabled day-night operations, while onboard processing filtered and prioritised imagery before transmission. When communication links were degraded, the system buffered critical data for later transmission or autonomous return. The AI-driven navigation algorithms continuously compared live sensor data against stored terrain profiles, sun angle calculations, and time-of-day models derived from the drone's last confirmed position. This allowed the aircraft to estimate its relative location with remarkable accuracy, even under sustained GPS denial, and to calculate an energy-efficient return-to-base trajectory before battery reserves reached critical thresholds.

Since its initial deployment, the platform has continued to evolve through a modular design philosophy that allows rapid adaptation to new missions and technologies. Variants have been developed using alternative propulsion units, higher-capacity energy storage systems, and refined aerodynamic surfaces to support extended-range operations. Some later iterations are reportedly capable of reaching distances of 400 to 500 kilometres behind enemy lines, pushing the boundaries of what would traditionally be expected from a hobby-derived air-frame. The modular construction approach—standardised fuselage sections, swappable wings, interchangeable avionics bays, and adaptable payload mounts—has enabled continuous iteration without the need for complete redesigns, dramatically shortening development cycles.

Perhaps most striking is the cost-to-capability ratio achieved by this system. Despite its sophisticated autonomy, sensor integration, and endurance performance, the drone remains based on an adapted hobby platform with a unit cost estimated between USD 2,000 and 3,000. When compared to functionally similar fixed-wing reconnaissance drones developed and fielded through NATO procurement channels—often costing between USD 140,000 and 160,000 per unit—the disparity is stark. While higher-end systems may offer greater redundancy, certification, and integration into legacy command structures, the Ukrainian example demonstrates how necessity-driven innovation, open-architecture design, and rapid field feedback can produce highly effective combat-ready systems at a fraction of the cost. In the context of modern high-attrition warfare, this approach has reshaped expectations of what small, fixed-wing unmanned aircraft can achieve, and has permanently altered the economics of aerial reconnaissance in contested environments.

Example 5 - Quad-Copter with Fibre Optic FPV Control in the Donbas, Oblast.

This photograph displays a high-speed quad-copter configured for long-range, high-risk strike missions, representing one of the most significant tactical evolution of the conflict in Ukraine toward the end of 2025.

The drone shown is equipped with an additional power source and an explosive payload mounted above the frame, optimised for one-way "kamikaze" missions extending between 40 and 60 kilometres into enemy territory. Its intended role is the precise destruction of medium-sized armoured vehicles or the disruption of ground-based platoon deployments, striking targets that were previously difficult or costly to reach using conventional means. Flown by a skilled pilot using first-person-view (FPV) goggles, the aircraft would typically be launched and controlled from a concealed position many kilometres behind the front lines, allowing the operator to exploit speed, agility, and terrain-hugging flight profiles to maximise surprise, impact, and psychological disruption against opposing Russian units.

One of the primary advantages of this class of drone lies in its fibre-optic control tether, a technological adaptation that has reshaped survivability in an increasingly contested electronic warfare environment. Unlike radio-controlled systems, which are vulnerable to jamming, spoofing, and signal interception, the fibre-optic link provides a near-immune command-and-control channel. This direct physical connection ensures uninterrupted, high-bandwidth transmission of control inputs and live video, even in areas saturated with electronic countermeasures. As a result, pilots retain precise control during the terminal phase of flight, where split-second adjustments can mean the difference between a successful strike and mission failure. This has dramatically increased hit probability against mobile or partially concealed targets, particularly armour operating under electronic protection umbrellas.

The tether also introduces secondary performance benefits that extend beyond communications resilience. In some configurations, the fibre-optic cable contributes a portion of the drone's power supply, supplementing onboard batteries and marginally extending endurance and operational reach. This hybrid power-control arrangement enables longer loiter times, more flexible routing, and greater tolerance for evasive manoeuvres at speed. Combined with lightweight frames, high-output motors, and aggressive flight tuning, these drones can fly fast, low, and unpredictably, making visual detection and kinetic interception extremely difficult. From a cost-effectiveness perspective, the disparity is stark: platforms costing only a few thousand dollars are capable of destroying vehicles and equipment worth orders of magnitude more, imposing disproportionate economic and logistical strain on the adversary.

Beyond physical damage, the psychological and tactical effects are equally significant. Fibre-optic FPV drones force opposing forces to assume constant vulnerability, even far behind nominal front lines. Their presence disrupts movement, complicates resupply, and erodes confidence in traditional defensive measures. Armoured units must disperse, conceal, or operate under restrictive conditions, reducing tempo and effectiveness. This shift has contributed to a broader transformation in combat dynamics, where small teams with relatively modest resources can generate outsized battlefield effects through speed, precision, and technical ingenuity.

However, despite their effectiveness, the widespread deployment of fibre-optic FPV drones has introduced a set of emerging and often overlooked challenges. Chief among these is the physical accumulation of fibre-optic cable across the battlefield. Thousands of missions leave behind kilometres of fine, nearly invisible strands draped across fields, forests, ruins, and urban environments. This creates a new class of environmental and operational hazard, entangling personnel, vehicles, and equipment, and complicating movement for both combatants and civilians. In wooded or urban areas, the cable can snag boots, weapons, or vehicle components, increasing the risk of injury or detection during ground manoeuvres.

These remnants also pose long-term post-conflict issues. Fibre-optic material does not biodegrade quickly and will require deliberate clearance efforts once hostilities end. Left unmanaged, it could interfere with agricultural recovery, reconstruction, and civilian mobility, effectively polluting the operational space long after the fighting has ceased. Additionally, the logistical burden of producing, transporting, and deploying large quantities of fibre-optic cable places strain on supply chains and increases preparation time for each mission. There are also technical trade-offs: the tether can limit extreme manoeuvres, introduce drag at longer distances, and create vulnerability if the cable is severed by terrain, debris, or defensive countermeasures.

In sum, fibre-optic FPV drones represent a profound shift in modern warfare, combining precision, resilience, and affordability in ways that have permanently altered battlefield realities. While their advantages in electronic warfare resistance, strike accuracy, and cost-effectiveness are undeniable, their continued use is simultaneously generating new environmental, logistical, and post-war challenges. As with many rapid wartime innovations, their legacy will extend beyond immediate tactical success, shaping both the future conduct of conflict and the complex task of recovery once the fighting ends.

Example 6 - Quad-Copter carrying a test payload (Thermite Payload).

The quad-copter illustrated in Example 6 represents the early stages of a development pathway that would become one of the most disruptive tactical innovations of the war. Fitted with a small thermite charge mounted beneath the air-frame, this class of drone emerged in the immediate aftermath of the full-scale Russian invasion in 2022, when Ukrainian drone units faced an urgent operational problem: how to disable heavily armoured vehicles operating behind enemy lines with limited access to conventional anti-tank weapons. Operating under the cover of darkness, confusion, or ongoing engagements, Ukrainian teams turned to hobby-grade quad-copters as an asymmetric solution capable of bypassing frontal armour entirely by attacking the most vulnerable part of any armoured platform—the engine compartment.

The principle behind this approach was brutally simple yet highly effective. A low-cost quad-copter, typically valued at approximately USD 1,200, could be guided onto the upper rear deck of a tank or armoured vehicle, where armour is thinnest and heat dissipation systems are exposed. Once positioned, the thermite charge would be ignited, burning at extremely high temperatures and rapidly melting through engine components, transmission housings, and critical support systems. Within seconds, the vehicle would be rendered immobile and combat ineffective, even if the crew survived. From a cost-exchange perspective, the results were overwhelming: a handful of inexpensive drones could permanently disable a vehicle worth several million dollars. Even when multiple drones were required to achieve success, the economic and operational imbalance strongly favoured Ukrainian forces.

As this tactic proved successful, the hobby drone units involved in its development quickly began working closely with professional military counterparts. This collaboration allowed access to specialised ordnance, improved fusing mechanisms, and refined deployment procedures, transforming improvised solutions into repeatable, scalable capabilities. Drone air-frames were adapted to reduce acoustic signatures through the use of lower-RPM propellers and vibration-damped mounts, often referred to by operators as "whisper" configurations. Infrared and thermal imaging cameras were integrated to enable precise positioning over heat-emitting engine blocks, allowing operators to identify and strike vehicles even in complete darkness or during periods of reduced visibility. Flight profiles were adjusted to prioritise low-altitude ingress, slow final approaches, and minimal lateral movement to avoid detection by both human observers and counter-drone sensors.

Over time, this style of attack did not stagnate but continued to evolve in response to Russian countermeasures. As electronic warfare became more prevalent, fibre-optic first-person-view (FPV) control systems were introduced, allowing drones to be flown deep into contested areas without reliance on vulnerable radio-frequency links. This innovation extended operational reach further into occupied territory and even across borders, while also reducing susceptibility to jamming and signal interception. The same principles were later adapted beyond land warfare, with quad-copters and FPV platforms being used in maritime environments near the Black Sea and around Odesa. Targets expanded to include repurposed oil platforms seized by Russian forces, logistics nodes, and naval vessels, demonstrating that neither land nor sea offered safe sanctuary.

What began in 2022 as an improvised response by small teams using hobby-grade equipment rapidly matured into a sophisticated and highly adaptive strike capability. The thermite-equipped or explosive quad-copter became emblematic of Ukraine's broader approach to drone warfare: low cost, rapid iteration, close integration between civilian innovators and military professionals, and an uncompromising focus on exploiting enemy vulnerabilities. This evolution underscores how necessity, combined with technical ingenuity and tactical discipline, reshaped the battlefield and permanently altered assumptions about the role of small unmanned systems in modern conflict.

Appendix (C) - Glossary of Terms

Glossary Abbreviation - Description of Term, abbreviation or reference (Table 1)

(AFE)	Analogue Front End	**(AI)**	Artificial Intelligence
(AIS)	Automatic Identification Systems	**(ADV)**	Ad-hoc On-demand Distance Vector
(AR)	Augmented Reality	**(Baud Rate)**	Speed of data transmission (1 baud = 1 bits per second)
(BMS)	Battery Management Systems	**(CFRPs)**	Carbon Fibre-Reinforced Polymers
(COTs)	Commercial off-the-Shelf	**(C-UAS)**	Counter UAS
(DIY)	Do it yourself	**(DoD)**	Depth of discharge
(DOSS)	Direct Sequence Spread Spectrum	**(ECM)**	Electronic Countermeasures
(EEZs)	Exclusive Economic Zones	**(ELRS)**	Is a hardware add-on for FPV remote controllers. Express LRS system to send low-latency, long-range control signals to drones
(EO)	Electro-Optical	**(ES/IR)**	Electro-Optical/Infrared
(ESC)	Electronic Speed Controller	**(ESCs)**	Electronic Stability Control
(ESR)	Equivalent Series Resistance	**(FC)**	Flight controller
(FHSS)	Frequency-Hopping Spread Spectrum	**(FMU)**	Flight Management Unit
(FPV)	First Person View	**(FRPs)**	Fibre Glass Reinforced Plastics
(GCPs)	Ground Control Points	**(GCS)**	Ground Control Station
(GHz)	A unit of frequency that describes the radio wave frequency used for communication.	**(GPEs)**	Gel Polymer Electrolytes
(GPS)	Global Positioning System	**(HMDs)**	Headmounted Displays
(HPM)	High-Power Microwave	**(ICE)**	Internal Combustion Engine
(IMUs)	Integrating Inertial Measurement Units	**(INS)**	Inertial Navigation Systems
(IR)	Internal resistance and or Infra Red context dependant	**(ISM bands)**	Industrial, Scientific, and Medical communication bands
(ISR)	Intelligence, Surveillance, and Reconnaissance	**(Li -PSB or Li-pSSB)**	Lithium Solid-State
(Li-Ion)	Lithium-Ion	**(Li-Po)**	Lithium Polymer
(Li-SSD)	Lithium Solid State	**(LOS)**	Line of Sight

Glossary Abbreviation - Description of Term, abbreviation or reference (Table 2)

(mAh)	Mills-Amp Hours, indicates the size of the battery's "fuel tank". A higher mAh number means a larger capacity and generally a longer runtime.	**(MANPADS)**	Man-Portable Air-Defense Systems
(MCU)	A Micro-Controller Unit	**(MOSFETs)**	Field-Effect Transistors, an Electronic Component
(MR)	Mixed-Reality	**(MUM-T)**	Manned-Unmanned Teaming
(mΩ)	(mΩ) - A measurement of electronic resistance, milliohms (mΩ)	**(NGOs)**	Non-Governmental Organisation, a non-profit group.
(NiMH)	Nickel-Metal Hydride	**(NMC)**	Nickel Manganese Cobalt
(OLSR)	Optimised Link State Routing	**(OSD)**	On Screen Display
(PEO)	Poly (ethylene oxide)	**(PID)**	Proportional-Integral-Derivative. A Mechanism in industrial
(PSYOPS)	Psychological Operations	**(PV)**	Process Variable
(RCS)	Radar Cross-Section	**(RF)**	Radio Frequency
(RPM)	Revs Per Minute	**(SAM)**	Surface-to-Air Missile
(SAR)	Synthetic Aperture Radar or Search and Rescue dependant on context.	**(SATCOM)**	Satellite Communications
(SEI)	Solid-Electrolyte Interphase	**(SoC)**	State of Charge
(SoH)	State of Health	**(SoP)**	State-of-Power
(SP)	Set point	**(UART)**	Universal Asynchronous Receiver/ Transmitter
(UAV RF)	Unmanned Aerial Vehicle (UAV) systems that utilize	**(UAVs)**	Unmanned Aerial Vehicles
(URLLC)	Ultra-Reliable Low-Latency Communications	**(V)**	Volts
(VARTM)	Vacuum-Assisted Resin Transfer Moulding	**(VR)**	Virtual Reality
(VTOL)	Vertical Take Off and Landing	**(VTX)**	Video Controller/Transmitter
(WDM)	Wavelength Division Multiplexing	**(Wh)**	Watt Hours
(Wh/Kg)	The amount of energy a battery can store relative to its weight	**(WiFi)**	WireLess Connections

Glossary Abbreviation - Description of Term, abbreviation or reference (Table 3)

Autonomous	Operate and complete tasks independently, without direct human control
(anode)	The negative electrode
BETAFlight Software	Betaflight is Open-Source flight controller software (firmware) for multi-rotor and fixed wing craft. If you're flying an FPV quadcopter, you're probably running Betaflight on the flight controller.
(cathode)	The Positive electrode
Cluster Munitions	Refers to weapons that release numerous smaller explosive "bomblets" over a wide area.
4G/5G	Generations of mobile (cellular) network technology. 4 Generation. Often called 4G LTE (Long Term Evolution) Fast mobile internet for everyday use, Video streaming (HD), Social media, web browsing, Video calls. Speed: ~10–100 Mbps (sometimes higher), Latency: ~30–50 ms. 5 Generation. Ultra-fast downloads (4K/8K streaming), very low delay (real-time control), connecting lots of devices at once, autonomous vehicles, industrial automation, drone control & live HD telemetry.Speed: 100 Mbps to 10+ Gbps, Latency: ~1–10 ms.
Firmware	Firmware is software that provides low-level control of computing device hardware
Fixed-Wing	A type of drone that looks and flies like a traditional air-plane, using rigid, stationary wings.
Hardware	Physical components of a computer system or in this context, electronic components and microprocessors Ina drone.
Hybrid	Something made by combining two or more different elements and/or systems.
Open-Source	Normally relates to software with publicly accessible source code that anyone can view, modify, and share.
Payload	The useful load a vehicle or system carries.
Radio-Controlled.	An object is operated or guided from a distance using invisible radio signals sent from a handheld transmitter
Software	Instructions written in code that tell a computer what to do, operate and function.
Swarming tactics	Using a large number of coordinated, often autonomous, units (like drones, soldiers, or vessels) to overwhelm an opponent.

Acknowledgements

This book was written in the shadow of a conflict that has fundamentally redefined modern warfare, but its origins were far more tentative. The work began during the author's first visit to Ukraine in 2022, at a time when the concept of artificial intelligence enabling truly autonomous combat drones was still widely regarded as speculative. While unmanned systems were already playing a significant role on the battlefield, they remained largely dependent on human pilots, with autonomy and AI-driven decision-making viewed as technologies of the future rather than operational realities.

Over the following three years of sustained work in and around Ukraine, that perception changed dramatically. What had once been theoretical rapidly became observable fact. Artificial intelligence began to play a direct and measurable role in combat drone operations, from AI-assisted navigation in GPS-denied environments to machine-learning systems capable of adapting flight behaviour, target recognition, and mission profiles under intense electronic warfare conditions. The progression from concept to capability unfolded in real time, transforming fiction into lived reality.

Since the Russian invasion in 2014, Ukraine has emerged as an unplanned yet extraordinary laboratory for the accelerated development of unmanned aerial systems. Early adaptations of civilian-grade drones for reconnaissance evolved into a complex and highly adaptive ecosystem of platforms integrating artificial intelligence, machine learning, encrypted and resilient communications, autonomous navigation, electronic-warfare resistance, and rapid battlefield-driven design iteration. Documenting this transformation presented both a significant technical challenge and a profound human responsibility.

First and foremost, recognition is owed to the people of Ukraine, whose resilience, ingenuity, and determination have driven this unprecedented technological evolution. Engineers, programmers, hobbyists, soldiers, medics, and volunteers, often working with limited resources and under constant threat, collectively reshaped how drones are designed, deployed, and understood. Their ability to adapt technology under fire has permanently altered global perspectives on unmanned systems and modern conflict.

The author also acknowledges the Ukrainian drone developers, operators, and innovators who, since 2014, consistently pushed beyond traditional military procurement cycles and rigid doctrine. Through necessity rather than theory, they developed modular systems, rapid-repair frameworks, open-architecture flight controllers, AI-assisted targeting and navigation, and resilient communication methods capable of surviving intense electronic warfare. The speed at which lessons were learned, shared, and redeployed across the battlefield stands without historical parallel.

Gratitude is extended to the legitimate humanitarian organisations, technical volunteers, and overseas supporters who worked alongside Ukraine throughout the conflict. Many contributed not only to humanitarian relief but also to the development, testing, transport, and ethical application of unmanned systems for reconnaissance, logistics, medical resupply, and civilian protection. Particular acknowledgement is given to UK Aid, DIY Ukraine, Ukraine Patriots Kyiv, Unbroken Lviv, UK Med, and the numerous hospitals and children's charities operating in Kyiv and Lviv under constant threat.

The author's friends and family are also acknowledged for their unwavering support throughout repeated deployments, prolonged periods of writing, and the emotional toll of observing a war where technology and humanity collide so starkly. Their patience and encouragement sustained the work from its earliest drafts to completion.

Finally, acknowledgement is given to the readers. This book is not solely about drones, algorithms, or hardware, but about adaptation under pressure, ethics in innovation, and the human cost behind every technological advance. If the work contributes even a modest understanding of how Ukraine's experience since 2014, particularly following 2022, has reshaped unmanned warfare, then it has fulfilled its purpose.

The book is dedicated to those who transformed necessity into innovation, and to the resilient people of Ukraine, whose endurance and adaptability will be studied for generations to come.

A special acknowledgement for the developers of www.BETAFlight.com, open source software.

Poem, war's moral rupture.

Poem, war's moral rupture, the industrialisation of killing, and the way drones erase distance, mercy, and humanity.

The Sky No Longer Looks Away

Once, men looked into each other's eyes and saw fear hesitate, saw mercy flicker like a faulty match.

Once, a hand could shake.
Once, a conscience had a face.
Now the sky hums.

Metal insects circle prayer and ruin alike, their shadows thinner than guilt, their patience infinite.

They do not sleep,.
They do not forget,
They do not bleed,

The battlefield has no horizon, only grids, coordinates, heat signatures, life reduced to colour on a screen far from the smell of blood.

A child runs,
The algorithm does not care why,
Distance has become a weapon, and courage a statistic,
Death arrives without hatred, without anger, without the dignity of being known,

Above the city, the drone waits, not hunting enemies, but seeking permission.

A click.,
A pulse,
A body disappears into data marked successful engagement,

No scream crosses the fibre,
No soul reaches the operator's chair,
The war is clean here,
The coffee is still warm,

Graves multiply faster than questions.

The machine learns faster than grief,
Each strike refines the next, the curve of a shoulder, the rhythm of a heartbeat, the lie that humanity can be mapped,

Soon, there is no enemy, only targets that failed to move fast enough.

History will call this progress,
Engineers will call it efficiency,
Generals will call it restraint,
The dead will call it nothing at all,

And when the last drone circles a silent, emptied world, still searching for warmth where none remains, the sky will finally look away, not in shame, but because there is no one left to watch it.

This book has been published as part of the series

Putin's Pathway series

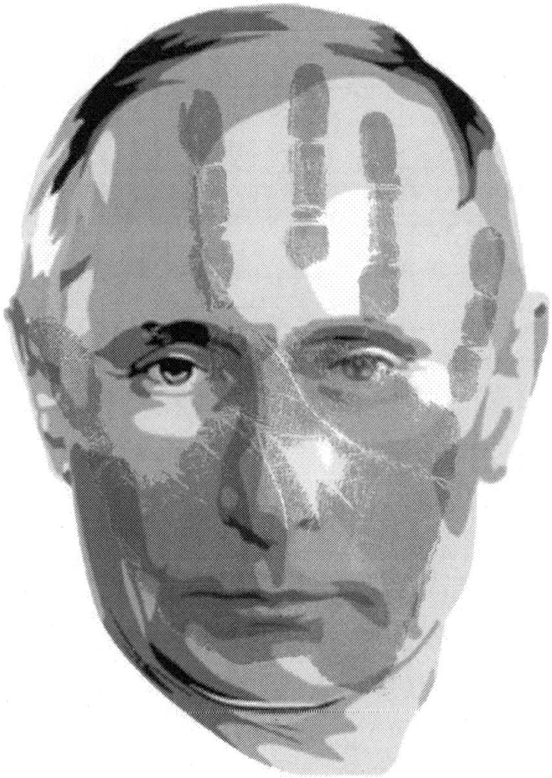

Made in the USA
Monee, IL
08 November 2026

68200535R10210